KIELMEYER AND THE ORGANIC WORLD

Also Available From Bloomsbury

The Schelling Reader, edited by Daniel Whistler and Benjamin Berger
The German Idealism Reader: Ideas, Responses, and Legacy,
edited by Marina F. Bykova
Deleuze and Ethology: A Philosophy of Entangled Life, Jason Cullen

KIELMEYER AND THE ORGANIC WORLD

Texts and Interpretations

Edited by
Lydia Azadpour and Daniel Whistler

BLOOMSBURY ACADEMIC
LONDON • NEW YORK • OXFORD • NEW DELHI • SYDNEY

BLOOMSBURY ACADEMIC
Bloomsbury Publishing Plc
50 Bedford Square, London, WC1B 3DP, UK
1385 Broadway, New York, NY 10018, USA
29 Earlsfort Terrace, Dublin 2, Ireland

BLOOMSBURY, BLOOMSBURY ACADEMIC and the Diana logo are trademarks of
Bloomsbury Publishing Plc

First published in Great Britain 2021
This paperback edition published in 2022

Copyright © Lydia Azadpour, Daniel Whistler, and Contributors, 2021

Lydia Azadpour and Daniel Whistler have asserted their right under the Copyright, Designs and Patents Act, 1988, to be identified as Editors of this work.

Cover design by Charlotte Daniels
Cover image © Getty Images

All rights reserved. No part of this publication may be reproduced or transmitted in any form or by any means, electronic or mechanical, including photocopying, recording, or any information storage or retrieval system, without prior permission in writing from the publishers.

Bloomsbury Publishing Plc does not have any control over, or responsibility for, any third-party websites referred to or in this book. All internet addresses given in this book were correct at the time of going to press. The author and publisher regret any inconvenience caused if addresses have changed or sites have ceased to exist, but can accept no responsibility for any such changes.

A catalogue record for this book is available from the British Library.

Library of Congress Cataloging-in-Publication Data

Names: Kielmeyer, Carl Friedrich, 1765-1844. Works. Selections. English. | Azadpour, Lydia, editor. | Whistler, Daniel, 1982- editor.
Title: Kielmeyer and the organic world : texts and interpretations / edited by Lydia Azadpour and Daniel Whistler.
Description: London ; New York : Bloomsbury Academic, 2021. | Includes bibliographical references and index.
Identifiers: LCCN 2020032068 (print) | LCCN 2020032069 (ebook) | ISBN 9781350143463 (hardback) | ISBN 9781350196711 (paperback) | ISBN 9781350143470 (ebook) | ISBN 9781350143487 (epub)
Subjects: LCSH: Kielmeyer, Carl Friedrich, 1765–1844. | Naturalists–Germany–Biography. | Philosophy of nature–Germany. | Biology–Philosophy.
Classification: LCC QH31.K53 K54 2021 (print) | LCC QH31.K53 (ebook) | DDC 508.092 [B]—dc23
LC record available at https://lccn.loc.gov/2020032068
LC ebook record available at https://lccn.loc.gov/2020032069

ISBN: HB: 978-1-3501-4346-3
PB: 978-1-3501-9671-1
ePDF: 978-1-3501-4347-0
eBook: 978-1-3501-4348-7

Typeset by RefineCatch Limited, Bungay, Suffolk

To find out more about our authors and books visit www.bloomsbury.com
and sign up for our newsletters.

CONTENTS

Acknowledgements	vii
Note on Contributors	viii

Chapter 1
EDITORS' INTRODUCTION 1
   Lydia Azadpour and Daniel Whistler

Chapter 2
KIELMEYER'S FAME AND FATE 11
   Kai Torsten Kanz

Part I
KIELMEYER'S 1793 SPEECH

Chapter 3
ON THE RELATIONS BETWEEN ORGANIC FORCES IN THE SERIES
OF DIFFERENT ORGANISATIONS, AND ON THE LAWS AND
CONSEQUENCES OF THESE RELATIONS 29
   C. F. Kielmeyer

Part II
SELECTED UNPUBLISHED TEXTS BY KIELMEYER

Chapter 4
ON NATURAL HISTORY 53
   C. F. Kielmeyer

Chapter 5
IDEAS FOR A DEVELOPMENTAL HISTORY OF THE EARTH AND ITS
ORGANISATIONS: LETTER TO WINDISCHMANN, 1804 65
   C. F. Kielmeyer

Chapter 6
ON KANT AND GERMAN PHILOSOPHY OF NATURE: LETTER
TO CUVIER, 1807 69
   C. F. Kielmeyer

Part III
INTERPRETATIONS

Chapter 7
FORCE AND LAW IN KIELMEYER'S 1793 SPEECH 81
   Andrew Cooper

Chapter 8
ORGANIC PHYSICS AS A PHENOMENOLOGY OF THE ORGANIC 99
   Thomas Bach

Chapter 9
THE PATH OF THE GREAT MACHINE: KIELMEYER'S ECONOMY OF
EXTINCTION 115
   Lydia Azadpour

Chapter 10
RECAPITULATION ALL THE WAY DOWN? PHILOSOPHICAL
ONTOGENY IN KIELMEYER AND SCHELLING 133
   Iain Hamilton Grant

Chapter 11
KIELMEYER AND THE CYBERNETICS OF THE ORGANIC WORLD 149
   Andrea Gambarotto

Chapter 12
REPRODUCTION, PRODUCTION AND THE EARTH: THE PLACE OF SEX IN
KIELMEYER'S 'ECONOMY OF THE ORGANIC WORLD' 171
   Susanne Lettow

Chapter 13
MECHANICS BEYOND THE MACHINE IN KIELMEYER
AND ESCHENMAYER 187
   Jocelyn Holland

Chapter 14
THE LOGIC OF ORGANIC FORCES: HEGEL'S CRITIQUE OF KIELMEYER 203
   Benjamin Berger

Bibliography 221
Index 233

ACKNOWLEDGEMENTS

First and foremost, we wish to thank Iain Grant for so generously making available his Kielmeyer translations for this volume; they have contributed so much to what is to follow. We are also very grateful to the contributors for all their hard work and to Liza Thompson, Lisa Goodrum and Lucy Russell at Bloomsbury for their support and advice. The British Society for the History of Philosophy and the British Academy kindly provided funds for events that were formative for this research. Christa Stadtler and Michael Jaworzyn helped improve the translations no end, and Paul Ziche and Henning Tegtmeyer helpfully looked over sections of a draft-translation of the 1793 Speech. We would also like to thank the staff and students of the Politics, International Relations and Philosophy department at Royal Holloway for creating a stimulating environment, especially Phoebe Page and Anthony Bruno for their feedback and discussion. Lydia would like to thank Dawn, Azad and Emily Azadpour, and Michael Jaworzyn for their encouragement. Daniel would like to thank Grace Whistler for her indispensable support throughout.

NOTE ON CONTRIBUTORS

Lydia Azadpour is a researcher at Royal Holloway, University of London. She is currently completing a project on the concept of species in the philosophies of nature of Kielmeyer, Hegel and Schelling, and has previously studied at the University of Warwick and KU Leuven, Belgium.

Thomas Bach received his PhD in Philosophy at the University of Stuttgart in 1999 with a dissertation on Kielmeyer and Schelling, subsequently published as *Biologie und Philosophie bei C. F. Kielmeyer und F. W. J. Schelling*. Since 2010, he has acted as research officer at Friedrich-Schiller-Universität Jena's Ernst-Haeckel-Haus, which he led temporarily from 2014–19. Since 2013, he has also led the team producing a new edition of Ernst Haeckel's correspondence.

Benjamin Berger is Visiting Assistant Professor of Philosophy at Kent State University. His research focuses on nineteenth-century idealism and the philosophy of nature. With Daniel Whistler, he is the author of *The Schelling-Eschenmayer Controversy, 1801* (Edinburgh University Press, 2020) and editor of *The Schelling Reader* (Bloomsbury, 2020).

Andrew Cooper is Assistant Professor of Philosophy at the University of Warwick. He is author of *The Tragedy of Philosophy: Kant's Critique of Judgment and the Project of Aesthetics* (2016) and is currently completing a manuscript on eighteenth-century natural history.

Andrea Gambarotto is a research fellow at the Institut Supérieur de Philosophie, UC Louvain. His research intertwines classical German philosophy, philosophy of nature and philosophy of biology. His driving question is whether the theoretical framework developed by German idealism can be reshaped into a philosophy of nature for the twenty-first century.

Iain Hamilton Grant is Senior Lecturer in Philosophy at the University of the West of England. He is the author of *Philosophies of Nature after Schelling* (Continuum, 2006), numerous articles on philosophy of nature and its contemporary implications and is the editor of the forthcoming edition of F.W.J. Schelling's *Writings on Philosophy of Nature* (SUNY Press).

Jocelyn Holland is Professor of Comparative Literature at Caltech. She has published three book-length projects: *German Romanticism and Science: The*

Procreative Poetics of Goethe, Novalis, and Ritter; *Key Texts by Johann Wilhelm Ritter on the Science and Art of Nature*; and *The Lever as Instrument of Reason: Technological Constructions of Knowledge around 1800*. Her current project investigates eighteenth-century theories of technology.

Kai Torsten Kanz is research fellow at the University library of Justus Liebig University Gießen, where he is preparing an online edition of the correspondence of the biologist Karl Ernst von Baer (1792–1876). His books include *Kielmeyer-Bibliographie* (Stuttgart, 1991), an edition of Kielmeyer's *Ueber die Verhältnisse der organischen Kräfte unter einander* (Marburg, 1993) and the monograph, *Nationalismus und internationale Zusammenarbeit in den Naturwissenschaften* (Stuttgart, 1997). He has edited *Philosophie des Organischen in der Goethezeit* (Stuttgart, 1994), and *Christian Gottfried Nees von Esenbeck: Briefwechsel mit Johann Wolfgang von Goethe nebst ergänzenden Schreiben* (Stuttgart, 2003).

Susanne Lettow is a senior researcher at the Margherita von Brentano Centre for Gender Studies at Freie Universität Berlin. She also teaches philosophy at Freie Universität Berlin and her research areas include feminist philosophy and theory, history and philosophy of the life sciences, biopolitics, social and political philosophy

Daniel Whistler is Reader in Modern European Philosophy at Royal Holloway, University of London. He is author of *Schelling's Theory of Symbolic Language: Forming the System of Identity* (Oxford University Press, 2013) and co-author of *The Schelling-Eschenmayer Controversy, 1801* (Edinburgh University Press, 2020), and has edited, among other volumes, *The Edinburgh Critical History of Nineteenth-Century Theology* (Edinburgh University Press, 2017).

Chapter 1

EDITORS' INTRODUCTION

Lydia Azadpour and Daniel Whistler

This is the first English-language volume dedicated to the work of Carl Friedrich Kielmeyer (1765–1844). It includes the first ever English translation of his famous 1793 Speech, *On the Relations of Organic Forces in the Series of Organisations*, in which he establishes nothing less than a conceptual framework for the understanding of the living natural world and its interactions. Further key texts by Kielmeyer are translated in Part Two of the volume and are followed, in Part Three, by a series of interpretations of his work that attempt to show its significance for both historical and contemporary reflections on nature.

In sum, these translations and commentaries aim to make sense of the recent resurgence of interest in Kielmeyer's project of conceptualizing organic nature among anglophone historians of philosophy and science, for whom he has increasingly come to represent not only the 'father of philosophy of nature', as Cuvier once described him[1], but also a key figure in the development of the life sciences in general. As long ago as Timothy Lenoir's *Strategies of Life* in 1982, Kielmeyer's contribution (alongside those of Kant and Kielmeyer's teacher, Blumenbach) to the 'teleomechanism' of early German biology had been appreciated[2]; however, it was not until the 2000s that Kielmeyer fully resurfaced in anglophone scholarship. One important landmark was Iain Hamilton Grant's influential *Philosophies of Nature after Schelling*, in which the significance of Kielmeyer's 'dynamic natural history'[3] was acknowledged not just in terms of its influence on F. W. J. Schelling and other philosophers of nature, but also in its own right. Indeed, according to Grant, Kielmeyer intervenes in transcendental philosophy to 'convert ... [Kant's] phenomenal and somatic nature ... into a priori dynamics.'[4]

If these are some of the recent historical precedents, then the years 2017 and 2018 marked a real watershed moment: two significant English-language works have appeared which are in large part founded on the demand to take Kielmeyer seriously. Hence – drawing on an earlier resurgence of German-language scholarship on Kielmeyer in the 1990s (one inaugurated by Kai Kanz' *Philosophie des Organischen in der Goethezeit* and continued by Thomas Bach's *Biologie und Philosophie bei C. F. Kielmeyer und F. W. J. Schelling*) – Andrea Gambarotto's *Vital*

Forces, Teleology and Organization subjected Lenoir's earlier work to scrutiny by focusing on Kielmeyer's mediating position between Blumenbach and many later developments in biology. The 1793 Speech, he writes, stands as 'the earliest *Systemprogramm* for biology as a unified science in Germany' insofar as it 'implements Blumenbach's framework and its lexicon but, instead of applying it to the individual organic body, applies it to all organic nature.'[5] Kielmeyer makes possible for the first time 'biology as a general, unified field – one concerned with the laws that regulate the organization of all living nature.'[6] Likewise, for John H. Zammito in *The Gestation of German Biology* (which appeared some months earlier), Kielmeyer's Speech contributes substantially to the creation of the scientific domain of biology, but such a founding gesture, nevertheless, goes far beyond the Göttingen connection to Blumenbach (still emphasised by Gambarotto) to ultimately encompass 'the essential impetus of the life sciences since Haller'[7], one 'that offered a systematic basis for the emergent science.'[8] Even more recently, Joan Steigerwald's 2019 *Experimenting at the Boundaries of Life* has furthered this anglophone Kielmeyer-renaissance by pointing to the importance of his 'cross-disciplinary ... explorations of the boundaries between organic and inorganic processes.'[9]

This volume constitutes the next step in this explosion of interest in Kielmeyer's biological programme and, in particular, furthers anglophone research on Kielmeyer in relation to three broad contexts: (a) in terms of the emergence of biology as 'a general, unified field' with its own distinct disciplinary identity; (b) in terms of his receptions in Idealism, Romanticism and the philosophy of nature; and (c) in terms of the intrinsic conceptual interest of his own thinking – that is, as a philosophical and scientific figure in his own right. The commentaries that make up Part Three of this volume contextualize him within the history of the life sciences, within philosophies of nature, Idealisms and Romanticisms; they discuss his relevance to contemporary theoretical perspectives on the organism, on ecology and on the social effects of developments in life science; and they analyse Kielmeyer's use of concepts of force, recapitulation, sexual difference and natural law. This volume, then, is an attempt to provide a comprehensive point of departure for future English-language research on Kielmeyer's historical and contemporary significance.

The Alpha and Omega of Philosophy of Nature

In order to contribute further to the growing case for why Kielmeyer matters, it is worth spending a moment on one example: his central importance to the evolution of comparative physiology at the turn of the nineteenth century in Idealism and philosophy of nature.

In 2001, Frederick Beiser inaugurated a decisive shift in the way in which English-language German Idealism scholarship approached the various attempts at the turn of the nineteenth century to philosophize about nature with the following programmatic statement:

Above the portals of the academy of absolute idealism there is written the inscription, 'Let no one enter who has not studied *Naturphilosophie*.' Without an understanding of at least the central doctrines, basic arguments and fundamental problems of *Naturphilosophie*, the absolute idealism of Schelling and Hegel is all but incomprehensible.[10]

Despite the compelling nature of this call to arms, it is nonetheless surprising that so few readers of German Idealism – Beiser himself, sometimes, included – have heeded it. Research on Hegel, Schelling and the early German Romantics still tends to be undertaken in isolation from serious study of the details of nature-philosophical doctrines. Beyond small pockets of Schelling scholarship, the domains of speculation, experimentation and analysis in the philosophy of nature remains unknown to many. Grant's 2006 claim still holds true: the writings of philosophers of nature remain 'largely unread'.[11]

Kielmeyer, however, can here act as a 'gateway thinker', for not only does he mediate between philosophy of nature, Kant and the life sciences in a genuinely transdisciplinary fashion, he is also a frequent source for later nature-philosophical thinking. It is Kielmeyer – rather than Kant, Leibniz or even Goethe – whose paternity of philosophy of nature is most frequently asserted, because it was his 1793 Speech (*On the Relations of Organic Forces in the Series of Organisations*) that took on the dual status as the first proper document of German philosophy of nature itself and as catalyst for much of the research into the organic that was to follow over the next twenty years. The Speech directly or indirectly inspired every major contribution to the philosophy of the organism in Germany at the turn of the nineteenth century – and its role as origin for the nature-philosophical tradition is encapsulated in Schelling's 1798 comment that it marked 'the advent of a new epoch in natural history'[12]. Nevertheless, it is also true that to reduce Kielmeyer *merely* to the role of precursor alone would be to ignore his continued – if relatively silent – presence over the subsequent decade. Indeed, one of the major frustrations for the later generation investigating the organic world was how little Kielmeyer went on to publish after 1793: his researches were made known only by way of the circulation of unofficial manuscripts, correspondence or the gossip of his Karlsschule and Tübingen students.[13]

His continued renown among philosophers of nature in the late 1790s can be gleaned from a debate that took place between Schelling and A.C.A. Eschenmayer at the turn of the century. In his *First Outline of the System of Philosophy of Nature*, Schelling had mentioned Kielmeyer among a list of philosophers capable of 'providing a *comparative physiology*' – 'a science not yet attempted', but which, when accomplished, would greatly further the cause of philosophy of nature.[14] And in one of his 1801 reviews of the *First Outline*, Eschenmayer – a former student of Kielmeyer at the Karlsschule – exploited this passing reference. He makes use of his insider knowledge of Kielmeyer's research to substantiate Schelling's hope and so portray Kielmeyer's 'second coming' – his return to the fray of academic publication – as the long-awaited messianic completion of the science of philosophy of nature. For Eschenmayer, the future of philosophy of nature depends on nothing less than the production of 'the first lines of a *physiologia comparata* of both plants and animals.' And he continues

> Perhaps Prof. *Kielmayer* [sic] in Tübingen might make known to us his results in these fields ... Unfortunately, this wealthy property owner provides so little assistance to our poor public funds. I mean here his invaluable *zoology*, a work in which he has equally invested his power and his time, two aspects which – in other minds not capable of subduing the natural law of inertia – stand typically in an inverse proportion. Furthermore, [this is] a work in which felicitous analogies and inductions and frequent intellectual insight (divination, as it were) have long been developed from the highest principles of philosophy of nature.[15]

According to Eschenmayer, therefore, Kielmeyer does not solely occupy the position of 'father' of philosophy of nature; rather, he is cast as the 'Messiah of nature-philosophical reason' as well. Kielmeyer is seen to both have begun the intellectual epoch and to promise its fulfilment in a comparative physiology. To his contemporaries, Kielmeyer's name represented hope for the salvific enrichment of their 'poor public funds'.[16]

Kielmeyer's Programme

Comparative Physiology

As Eschenmayer and Schelling appreciated, the comparative work that Kielmeyer was engaged in was not confined to formal anatomical comparisons, but also looked at anatomical differences as manifestations of proportions of organic force. This, for Kielmeyer, amounts to a focus on *function* in his understanding of organic force, and thus on a *comparative physiology*, rather than a morphology.[17]

In the 1793 Speech in particular, identifying organic forces and drawing them into a comparative physiology was a central aim. Kielmeyer's goal is to explain the composition and development of organic nature through the distribution of proportions of the organic forces that he identified: sensibility, irritability, reproduction, propulsion and secretion. His aim was to reveal regularities in the ways these forces combined within individual organisations, in species, and in the whole of the organic world or nature, such that he could discern laws governing the dynamics of these proportions. His resulting speculation was that animal kinds could be arranged in a series according to the proportions of force they manifest. According to Kielmeyer's hierarchical series, 'higher' animals demonstrated a greater share of sensibility, whose share is increasingly restricted as Kielmeyer's chain of being is descended. As the share of sensibility is decreased, though, reproduction is increasingly manifest. As Bersier and others have discussed, Kielmeyer conceives of this via a 'law of compensation', where a lack in one force in a species is offset by a wealth in another.[18]

What does it mean to say that Kielmeyer was interested in the 'organic' world – as the title of this volume suggests? For Kielmeyer, the 'great machine of the organic world' was comprised of a series of 'organisations'.[19] At this time, the emphasis on organisation would have suggested not only a particular structure[20] of the phenomenon under consideration, but also that this phenomenon itself was the result of a process or activity. These features of 'organisations' – as structure that is activity or process – can be seen clearly in Kielmeyer's introduction of the description of the interrelations of organs in a system. An organisation is spatiotemporally 'animated by its organs' in such a way that 'each organ is so adapted to the changes of all other organs and so united in a system of simultaneous and consecutive changes that according to our manner of speaking, each is reciprocally cause and effect of the other.' (p. 30)[21] Kielmeyer's description of the 'organic world' as a larger whole shares such a description of its structural and processual character with that belonging to particular organisations, where an organisation is 1) a certain kind of structure, and 2) a structure that is the result of ongoing processes or activities. This claim about organs in an organism – that each is 'reciprocally cause and effect of the other' – of course echoes Kant's discussion of organized beings in the *Critique of Judgment*, which claims that 'we must think of each part [of a product of nature] as an organ that produces the other parts (so that each reciprocally produces the other)'.[22] That an organisation seems to manifest mutual determination is only possible given the organic forces Kielmeyer identifies, because they supply this simultaneity of means and ends by way of combining within an organism.

Method and Ontology of Nature

When we study nature, Kielmeyer claims, the object of study is often not clearly defined (pp. 53–4). The term 'nature' has indeed been applied very broadly, he claims – more and more so over time. This has happened particularly to 'nature', because the concept includes at its root the idea of emergence, and 'emergence' is extended to describe the cause of a change, a historical presentation of the phenomenon, denoting a kind, or the laws governing some change. Because of the proximity of cause and effect, the term can also denote *natura naturans, natura naturata*, physiological laws, laws grounded in a highest cause, something's essence, or the sum of God's properties (pp. 54–5).

Given this proliferation of possible meanings, Kielmeyer isolates what he thinks the natural scientist proper must mean by 'nature':

> Nature is everything emergent or actually appearing to our senses and which is perceived with our inner and outer sense, connected in space and time, and apt to follow certain laws.
>
> p. 55

His decision to define nature in this way, i.e. by focusing on its phenomenal appearance – rather than on some of the other possible, more ontological meanings he enumerates – suggests a refusal to develop robust ontological theses. And unsurprisingly, throughout his work, the ontological status of the forces Kielmeyer identifies is unclear – perhaps deliberately so. Many of Kielmeyer's phrases suggest a provisional attitude to his hypotheses, and his discussion of force is no exception: he writes, for example, that his use of the concept of force is 'makeshift' (p. 32) and that his schema of five forces could be 'cancelled by a higher understanding' (p. 32). Throughout, he remains open to the possibility that we may perceive differences where there are in fact expressions of a unified power of some kind. To Kielmeyer in 1793, it seemed acceptable for the time being – and even 'conducive' (p. 31) – to postulate hypotheses in a provisional way.

Kielmeyer was undoubtedly influenced by Blumenbach, with whom he spent time in Göttingen: both of their projects seek to discern the vital forces at work in organisations. But while Blumenbach held that his formative force was 'not referable to any qualities merely physical, chemical, or mechanical'[23], it is not the case that Kielmeyer believed that his organic forces were ultimately unconnected to mechanical or chemical ones. As his claim about the 'makeshift' character of his forces shows, care must be taken not to hypostatize the organic forces in Kielmeyer's investigations. We can see this not only in the provisional, hesitant rhetoric of his writing, but also in the content of his works. When discussing the reproductive force, he writes that it *could* be seen as a distinguishing feature of organic beings if it were not that 'as in previous cases [it could] be sufficiently demonstrated, that it too sprung from inorganic nature and can be derived from forces internal [to the inorganic realm]' (p. 38). This hesitation leads to all sorts of scholarly debates: Lefèvre and Klein claim that Kielmeyer, 'in principle, also believed in a chemical explanation of the processes of life' (in his early work, at least)[24]; however, according to Lenoir, he did not yet think that chemistry had reached a stage at which it would be able to undertake this task.[25]

The History of Nature

As evidenced by the manuscripts and lectures in this volume, Kielmeyer held that investigating nature's past is critical to properly understanding it. This held not only for geological change, which was widely accepted, but for living phenomena too. That we must ask not only how nature *is*, but how it *was* and *will be* distinguishes natural history proper, for Kielmeyer, from mere description. This commitment from his early manuscripts (p. 60) is carried over into his concern to discuss the course of animate nature in his 1793 Speech and into his later discussions of the possibility of species extinction in his correspondence with Cuvier.

In addition, Kielmeyer is often acknowledged as a forerunner of recapitulation (the assertion that ontogeny recapitulates phylogeny), due to his claim that the distribution of proportions of force across the series of organisations follows the same order as their distribution within a single individual's development. Given this, Kielmeyer adds, 'one could be led by carefully selected analogies to [the point of] assuming such a material cause as an explanation of the developmental appearances that could be imagined as at work in the first production of organisations on our Earth' (p. 43). However, as Gliboff points out, such parallels between embryonic development and development of species were 'not morphological', but rather 'kindred developmental processes, governed by the same laws and forces, which became active in similar temporal patterns.'[26]

The Structure of this Volume

The translations and interpretative essays that make up *Kielmeyer and the Organic World* are arranged as follows. This Introduction is followed by a detailed intellectual biography of Kielmeyer by Kai Torsten Kanz that thoroughly contextualizes the accompanying translations in terms of his scientific context and personal trajectory. Kanz argues that it is only through such a a close study of Kielmeyer's biography – based on his unpublished correspondence and papers – that a more balanced view of his place in the history of science will finally emerge.

Parts One and Two consist of a series of translations. Part One reproduces the 1793 Speech on the relations between organic forces. It was this speech, of course, that proved to be a catalyst for much of the research in philosophy of nature that came after Kielmeyer and that, therefore, stands as one of the key texts in the genesis of philosophies of nature in German Idealism and Romanticism, as well as a monument of reflection on the principles of the eighteenth-century life sciences. Short, unpublished texts in Part Two supplement the 1793 Speech by giving further evidence of Kielmeyer's thinking about nature, natural history, geological formation, the earth and contemporary philosophy. The lecture notes from 1790 on natural history provide a fragmentary insight into his broader conceptions of the possibility, remit and method of scientific inquiry into the organic world; the letter to Windischmann from 1794 shows Kielmeyer's sober assessment of the attempts to draw geological investigation and the life sciences into conversation and thereby sheds light on the 'global' implications of his research programme; and, finally, the letter to Cuvier from 1809 is Kielmeyer's most explicit and sustained intervention into the philosophical developments of his time and comprises a number of rigorous criticisms of the Kantian and post-Kantian intellectual landscape that had come to dominate the German world of letters.

Part Three comprises a number of commentaries on the meaning of Kielmeyer's 1793 Speech, on its place in his thinking as a whole as well as in the context of contemporaneous developments in the life sciences and philosophy, and also on the enduring significance of its conceptual framework. In Chapter Seven, Andrew Cooper reads the 1793 Speech as an exemplary instance of 'analogical Newtonianism', transposing the agnostic forms of reasoning authorized by Newton into the study of living beings. Thomas Bach also focus on the intellectual traditions in which Kielmeyer's project is to be situated, switching from the Newtonian context to the extension of 'physics' into the organic realm in Germany at the end of the eighteenth century. For Bach, Kielmeyer's Speech furthers a phenomenological programme of describing the effects of forces without determining precisely their causes.

Chapter Nine and Ten shift to specific Kielmeyerian concepts. Lydia Azadpour undertakes a detailed reading of the 1793 Speech to explain Kielmeyer's account of the history of the organic world in terms of his use of the concept of equilibrium. Her account shows how, in contrast to earlier 'natural economies', Kielmeyer's view of the equilibrium of organic forces allows him to see genuinely historical changes in nature – such as extinctions – as part of the internal dynamic of the organic world. Iain Hamilton Grant then moves attention to the concept of recapitulation implicitly formulated in the speech and elsewhere in Kielmeyer's corpus and subsequently developed by Schelling. Looking to the letter to Windischmann in particular, Grant discerns recapitulation at play in not only biological, but also geological and noological phenomena, suggesting that much more work needs to be undertaken on the temporality and historicity of Kielmeyerian nature.

In the subsequent two chapters, Andrea Gambarotto and Susanne Lettow begin to consider the implications and significance of Kielmeyer's research programme for concepts of teleology and sex difference, respectively. Gambarotto places Kielmeyer's remarks on intention and purpose in a tradition that leads into mid-twentieth-century cybernetic debate over self-organisation, and Lettow shows the intersections and interactions that emerge out of Kielmeyer's work into a whole host of idealist and nature-philosophical discourses on sex difference in the natural world. Of course, by making Kielmeyer available to an English-speaking audience, and by engaging in thematic analyses of Kielmeyer's work, what follows also invites further investigations into varied fields of influence, such as Kielmeyer's relation to the history of racism and the idea of race, particularly given his proximity to Blumenbach and Cuvier who are deeply involved in that history.[27]

In the final two chapters, attention shifts to Kielmeyer's legacy. In Chapter Thirteen, Jocelyn Holland reconstructs Kielmeyer's 1799 response to the work of his former student at the Karlsschule, Eschenmayer. This dialogue, she demonstrates, is not merely of interest for understanding criticisms of Eschenmayer's 'nature-metaphysics', but also sheds light on the relation between mechanism and organicism – between the images of the 'clock' and the 'organism', as Dorothea Kuhn once put it[28] – at the heart of Kielmeyer's thought. Finally, in Chapter Fourteen, Benjamin Berger focuses on one of the many responses to Kielmeyer from the Idealist tradition – Hegel's criticism of the 1793 Speech in both the *Phenomenology of Spirit* and the sections on philosophy of nature from the *Encyclopaedia*. Berger shows that at the heart of the Hegel-Kielmeyer constellation is a question over logical versus historical understandings of the natural world..

In short, therefore, the translations and commentaries that follow attempt to provide a comprehensive set of provocations that both introduce Kielmeyer to an English-language readership in a more sustained way than has previously been possible and stimulate further debate on the meaning, implications and significance of his writings.

Notes

1 See the sustained discussion of Cuvier's claim in T. Bach, 'Kielmeyer als "Vater der Naturphilosophie?" Anmerkungen zu seiner Rezeption im deutschen Idealismus', in

K. T. Kanz (ed.), *Philosophie des Organischen in der Goethezeit: Studien zu Werk und Wirkung des Naturforschers Carl Friedrich Kielmeyer (1765–1844)* (Stuttgart: Franz Steiner, 1994).
2   T. Lenoir, *The Strategy of Life: Teleology and Mechanics in Nineteenth-Century Biology* (Dordrecht: Reidel, 1982).
3   I. H. Grant, *Philosophies of Nature after Schelling* (London: Continuum, 2006), 49.
4   Ibid., 137.
5   A. Gambarotto, *Vital Forces, Teleology, and Organization: Philosophy of Nature and the Rise of Biology in Germany* (Dordrecht: Springer, 2018), 42–3.
6   Ibid., 46.
7   J. H. Zammito, *The Gestation of German Biology: Philosophy and Physiology from Stahl to Schelling* (Chicago: University of Chicago Press, 2018), 251.
8   Ibid., 3.
9   J. Steigerwald, *Experimenting at the Boundaries of Life: Organic Vitality in Germany around 1800* (Pittsburgh: University of Pittsburgh Press, 2019), 196.
10 F. C. Beiser, *German Idealism: The Struggle against Subjectivism, 1781–1801* (Harvard: Harvard University Press, 2002), 506–7.
11 Grant, *Philosophies of Nature*, ix.
12 F. W. J. Schelling, *Von der Weltseele – Eine Hypothese der Höhern Physik zur Erklärung des allgemeinen Organismus* (Stuttgart: Frommann-Holzboog, 2000), 201.
13 See, for example, S. Lettow, 'Generation, Genealogy, and Time' in Lettow, *Reproduction*, 31 and T. Bach, *Biologie und Philosophie bei C.F. Kielmeyer und F.W.J. Schelling* (Stuttgart: Frommann-Holzboog, 2001), 63. Owing to the fact that Kielmeyer published little besides his 1793 Speech, it is to his lectures that we must turn to gain some idea of his intellectual occupations and the key problems he was engaged with: besides teaching chemistry, he lectured on zoology, comparative anatomy and plant physics – for more details, see Ingrid Schumacher's outline of the courses Kielmeyer was teaching at the Karlsschule (in 'Karl Friedrich Kielmeyer, ein Wegbereiter neuer Ideen', in *Medizinhistorisches Journal* 14.1/2 [1979], 81–99).
14 F. W. J. Schelling, *First Outline of a System of the Philosophy of Nature*, trans. Keith R. Peterson (Albany, NY: SUNY Press, 2001), 50.
15 A. K. A. Eschenmayer, 'Spontaneity = World Soul, or On the Highest Principle of Philosophy of Nature', trans. J. Kahl and D. Whistler, in B. Berger and D. Whistler, *The Schelling-Eschenmayer Controversy, 1801: Nature and Identity* (Edinburgh: Edinburgh University Press, 2020), 44.
16 On the messianic structure of philosophy of nature alluded to in the above, see further D. Whistler, 'In the Hope of a Philosopher of Nature', in A. Ezekiel and K. Mihaylova (eds), *Hope and the Limits of the Self in Classical German Philosophy* (Berlin: De Gruyter, 2021 forthcoming).
17 Thus, 'irritability' denotes various perceivable effects, such as a reflex motion in a muscle fibre. However, it also denotes the underlying, unperceived cause of these effects (for example, what excites certain muscles to move the limb). When describing each force, Kielmeyer writes that each is defined as a certain 'capacity', and, he uses this term not only to refer to potencies and their actualization, but also sometimes to refer to the organ that possesses the power.
18 See, e.g. G. Bersier, 'Visualising Carl Friedrich Kielmeyer's Organic Forces: Goethe's Morphology on the Threshold of Evolution.' *Monatshefte* 97.1 (2005).
19 While Kielmeyer tends to discuss 'organisations', rather than 'organisms' in his 1793 Speech, he does also use the term 'organism'. The latter is used to denote (1) that the

organism *of* a living individual is made up of a system of organs; (2) that the 'organism' is the bearer of systems, such as irritability and sensibility; and (3) that the 'organism' is that in which systems or kinds of force are united. For Kielmeyer, 'organism' at least emphasises a key aspect of living *individuals* (i.e. that they are composed of systems). Cheung has credited Kielmeyer with being one of the first to 'systematically use the term "organism" as a generic name for individual entities' – a new use of the term that is 'focused on the "material individuality" of the "organism" as a specific form of Dasein and Organisation' (T. Cheung 'From the Organism of a Body to the Body of an Organism: Occurrence and Meaning of the Word "Organism" from the Seventeenth to the Nineteenth Centuries' in *The British Journal for the History of Science* 39.3 [2006], 330). For the most part, however, Kielmeyer discusses 'organisations' (always in plural) – various things, living beings, which are 'organised'.

20 See e.g. the entry for 'die Organisation' in *Deutsches Wörterbuch von Jacob und Wilhelm Grimm* (Leipzig 1854–1961).
21 Throughout, citations to Kielmeyer's texts translated in this volume are given in the form of in-text page numbers referring the reader to our translations themselves. Other references to Kielmeyer's writings taken from his *Gesammelte Schriften* (*Natur und Kraft. Gesammelte Schriften*, ed. F.-H. Holler [Berlin: Keiper, 1938]) are cited in-text using the abbreviation *GS* in what follows.
22 I. Kant, *Critique of Judgment*, trans. W. S Pluhar (Indianapolis, IN: Hackett, 2010), §65: 253.
23 J. F. Blumenbach, *Über den Bildungstrieb* (Göttingen: Dieterich, 1791), 22.
24 U. Klein and W. Lefèvre, *Materials in Eighteenth-Century Science: A Historical Ontology* (Cambridge, MA: MIT Press, 2007), 252. Klein and Lefèvre point out that, later in the 1800s, he thought that the elements constituting an organism 'could only be destroyed, but not created, by chemical art, since chemical art is not in command of the forces necessary for their resynthesis'.
25 Lenoir writes that Kielmeyer 'conceded that the French chemists had made advances in the chemical analysis of plants and animals. But he added that the chemical analysis of organic materials was still in its infancy, and further that no satisfactory application of chemical methods to the general theory of organisation could be expected in the near future.' (*Strategy of Life*, 64).
26 S. Gliboff, *H.G. Bronn, Ernst Haeckel, and the Origins of German Darwinism* (Cambridge, MA: MIT Press, 2008), 45.
27 Discussions of this issue in the period can be found in Suzanne Lettow's 'Introduction' in Suzanne Lettow eds, *Reproduction, Race and Gender*, (Albany: SU NY Press, 2014) and in P. H. Reill's *Vitalizing Nature the Enlightenment* (Berkeley: University of California Press, 2005), 199–236.
28 D. Kuhn, 'Uhrwerk oder Organismus. Karl Friedrich Kielmeyers System der organischen Kräfte', in *Nova Acta Leopoldina*, Neue Folge 36 (1970).

Chapter 2

KIELMEYER'S FAME AND FATE

Kai Torsten Kanz

Introduction

Today more than ever, the natural scientist Carl Friedrich von Kielmeyer (1765–1844) has become a subject of interest for historians of science, over 250 years after his birth. This recent growth of interest in his work and its impact is manifest in essays and book chapters on Kielmeyer, as well as in the translation of his major work, first into French[1] and now into English. This widespread reappreciation of his work raises the question: how did a scientist so famous in his own time go unrecognized for so long? That is, it raises the question: how could one of the key figures in medicine, natural science and philosophy at the turn of the nineteenth century have been forgotten for such a long time?

In the following chapter, I will describe Kielmeyer's renown during his own lifetime as well as the impact of his work on his contemporaries and on the history of science. I will argue that the key to understanding the complexity of how he has been viewed throughout history lies in Kielmeyer's biography. Both his contemporaries and later historians credit him with a quite specific role in the history of science, whether as a teacher, as a founder of a certain discipline, as the representative of a new scientific methodology, or even as the father of philosophy of nature in the Romantic era. Most of these, in part, contradictory and, in many cases, unfounded claims were already in the air during Kielmeyer's lifetime. In recent years, some of them have again been the cause of intense controversy. When looking at the significance of his work and its impact, it is necessary to critically interrogate the warrant for such claims.

Aside from the effect of his work, Kielmeyer himself formed part of an intellectual tradition that has not yet been explored in proper detail. Standing between the Age of Enlightenment and the Romantic period – and at the same time anticipating many developments from the nineteenth century – his work is both a reflection of his period but also timeless. Its focus lies between medicine, a subject Kielmeyer himself studied, and foundational natural science across many diverse research fields. Owing to the incredible philosophical impact his ideas had, it is not easy to intellectually position or contextualize Kielmeyer. He is described by some as a 'forgotten genius'[2], so as to make the question of his intellectual influence all but irrelevant. Others, however, reduce the originality of his thinking to his education at the Hohe Karlsschule in Stuttgart or his years of study at the University of Göttingen. Kielmeyer, it is certain, was significantly influenced by his professors at both these distinguished institutions, which were so influential on the history of German education. It is too simple just to see him as another renowned student of the Karlsschule alongside the more famous Friedrich Schiller, Georges Cuvier or Johann Heinrich Dannecker. But it is equally unsatisfying to view him as one of the intellectuals associated with the Göttingen professor Johann Friedrich Blumenbach. There are persistent images of Kielmeyer as just a 'disciple of Blumenbach's', as well as 'Cuvier's teacher'. And such images do have certain advantages; indeed, this gesture of turning Kielmeyer into a transitional figure on the threshold of two significant epochs is even to be found in the picture presented by one of his students a hundred years after his birth. According to the anatomist August Mayer in Bonn[3], Kielmeyer's lectures 'fill the great gap both in the history of comparative physiology from Blumenbach to Cuvier and in the history of chemistry from Fourcroy to H. Davy'. Yet, so positioned, Kielmeyer's debt to the preceding era as well as his impact on the subsequent era still remains indeterminate.

Kielmeyer's Education

To begin, I want to briefly outline Kielmeyer's scientific education, so as to identify some of the influences on him and his work, beginning with the Karlsschule in Stuttgart, his studies in Göttingen, his tours through Germany, his professorships in Stuttgart and Tübingen, and ending with his work as privy counsellor in Stuttgart.

Born on 22 October 1765 in Bebenhausen near Tübingen, as the son of a ducal keeper of hunting equipment, Kielmeyer was admitted to the Karlsschule in Stuttgart in 1773. After a period of philosophical instruction, he went on to study medicine. At the Hohe Karlsschule, natural history was taught by Carl Heinrich Köstlin, an inspiring young researcher. Whilst on a trip to Italy, Köstlin had met Alessandro Volta and he went on to translate Volta's work on the inflammable air of swamps into German. Köstlin, who taught there from 1780 until his premature death in 1783, used Blumenbach's *Handbuch der Naturgeschichte* as his textbook,[4] and, in fact, at this time, almost all the science teachers at the Karlsschule were using the popular textbooks penned by the professors at the University of Göttingen.[5]

Even more important than these references to his teachers and their choice of textbooks is Kielmeyer's self-education in chemistry and physiology described in his autobiographical sketch. Such autodidactic study was necessary because, for reasons unknown to us, he was forbidden by the director of the Hohe Karlsschule from attending the chemistry lectures. Moreover, when confronted with a lack of useful lessons in physiology, he again overcame such a difficulty through his own study. At this time, he primarily attended lectures on practical medical subjects;

yet, it was these two self-taught subjects that would go on to become Kielmeyer's central fields of research and teaching: 'He taught himself those aspects of medicine, which, like physiology, were later the focus of his work' (GS, 8). Surprisingly, considering his lack of formal education in the subject, Kielmeyer completed his medical studies in 1786 with a chemistry dissertation on the analysis of the mineral waters from Göppingen and Stuttgart-Berg.

Following his degree, Kielmeyer spent nearly a year and a half – from 22 December 1786 until 1 June 1788 – in Göttingen, supported financially by Duke Carl Eugen. During this time, he encountered a group of fellow students from Stuttgart, including his friend Johann Friedrich Pfaff. In Göttingen, he attended lectures by Georg Christoph Lichtenberg, Blumenbach, Johann Friedrich Gmelin and Abraham Gotthelf Kästner. His choice clearly demonstrates that he was not concerned with deepening his medical knowledge as part of his training, rather, his focus was on fundamental subjects of natural science, such as astronomy, physics and chemistry. He attended Blumenbach's lectures on animal anatomy (*Thieranatomie*) and natural history (*Naturgeschichte*), but the main reason for his acquaintanceship with Blumenbach was to gain access to the professor's famous natural history cabinet.[6]

Kielmeyer developed a much closer connection with Lichtenberg than with Blumenbach: Kielmeyer regularly visited and corresponded with Lichtenberg. Lichtenberg declared when discussing the solution of a scientific problem: 'I have never seen such a well-developed spirit of observation in someone of his age. I have the greatest expectations of him'.[7] The illustrious reputation of the University of Göttingen lay not only in the publications and lectures of its professors but also in the accessibility and size of its library. Kielmeyer was a frequent user of this rich source of literature. The lending records of the university library precisely document his reading matter at that time.[8] Moreover, he made use of the Göttingen inventory while preparing a manuscript on the sixteenth-century botanical reformers in Württemberg, which was edited posthumously (GS, 257–73).

After his studies in Göttingen, Kielmeyer toured through Germany, a trip that took him via Helmstedt and Magdeburg to Berlin and Potsdam. The return journey led him to Dresden, Freiberg, Halle, Jena and Erlangen, allowing Kielmeyer to visit the leading universities of the time. On this educational trip he met famous and less well-known academics – the latter group being preferred, since they were 'not so preoccupied with their own immortality'. He particularly mentions Johann Reinhold Forster, Marcus Elieser Bloch, Martin Heinrich Klaproth and August Johann Georg Karl Batsch (GS, 10).

On 1 February 1790, Kielmeyer was appointed Lecturer (*Lehrer*) of Zoology at the Hohe Karlsschule. At first, for his lectures on animal history (*Thier-Geschichte*) he used Blumenbach's *Handbuch der Naturgeschichte*, of which a third edition had just appeared. However, from 1791 onwards, he read from his own papers; and so, crucially, from this point onwards he based all his lectures on his own findings.[9] Two years later, in 1792, he became Professor of Chemistry and a full member of the medical faculty. On 11 February 1793, in celebration of the sixty-fifth birthday of Duke Carl Eugen von Württemberg, he gave the speech which would make him famous: *Ueber die Verhältniße der organischen Kräfte unter einander in der Reihe der verschiedenen Organisationen, die Geseze und Folgen dieser Verhältniße.*[10] This short work, which, when printed, ran to just forty-six pages, represents the moment in his career at which Kielmeyer began to be publicly recognized.

Freed from his teaching duties following the dissolution of the Hohe Karlsschule in Easter 1794, Kielmeyer undertook a second scientific journey that led him once again to Göttingen. From there he travelled to the North Sea, the Baltic Sea and to Heligoland where he spent time undertaking marine-zoological research. In 1796, he was transferred to the University of Tübingen with a full professorship in chemistry. Five years later, in 1801, he entered the medical faculty where he taught, additionally, botany and materia medica. Here he remained until 1817, productive and inspiring many, yet publishing next to nothing.

Kielmeyer's Lectures

Kielmeyer gave lectures at the Hohe Karlsschule in Stuttgart between 1790 and 1794 and at the University of Tübingen from 1796 to 1817. These lectures, covering a wide range of medical and scientific subjects, were primarily intended for students of medicine but were also attended by those from other faculties.

To begin, Kielmeyer's lectures at Stuttgart (1790–4) focused on zoology, but, from 1792 also covered chemistry. They were attended by the botanist Carl Friedrich von Gärtner, the physician Johann Heinrich von Autenrieth, the physicist Christoph Heinrich Pfaff and the psychologist Karl August Eschenmayer. These men – except Gärtner, who was a private scholar – became renowned professors and all referred to Kielmeyer's lectures in their later publications. Pfaff's dissertation on Galvanism in 1793 created a furore and was highly praised by Alexander von Humboldt and others[11], and it was Pfaff who was responsible for sending Georges Cuvier in France his notes from Kielmeyer's zoology lectures. This raises the question of whether Cuvier in fact adopted Kielmeyer's ideas for the purpose of his comparative anatomy and his law of correlation (then used as the basis for the foundation of vertebrate palaeontology). It is certainly true that, when Cuvier was a student in 1785/86, Kielmeyer introduced him to the art of animal dissection, and also that he was often thereafter described as 'Kielmeyer's pupil'.

In Tübingen between 1796 and 1817 following his appointment as a member of the medical faculty, Kielmeyer supervised the doctoral theses of thirty-two students.[12] His well-known postgraduates include the physicians Carl Eberhard Schelling (the brother of the philosopher Schelling) and Friedrich Schnurrer, the botanists Johannes Hegetschweiler and Ernst Gottlieb Steudel, the physician and palaeontologist Georg Friedrich Jäger, and the future Tübingen professors Ferdinand Gottlob Gmelin, Gustav Schübler and Wilhelm Ludwig Rapp. Numerous others, of whom he was not the tutor, also belonged to his circle of students, e.g. the chemist Leopold Gmelin in Heidelberg, the physician and poet Justinus Kerner in Weinsberg, and the anatomist August Karl Joseph Mayer in Bonn.

In spite of announcing their forthcoming publication, Kielmeyer's lectures remained unprinted during the decades of his teaching activities for unknown reasons. And it is because of their unavailability that his Tübingen lectures became exclusive events, with students travelling from near and far to hear the much-praised professor. Evidence from students who had first studied at Halle and then in Tübingen document the great attraction of the Tübingen medical faculty in the first decades of the nineteenth century as well as its supremacy over that of Halle.[13] This was primarily due to the pre-eminence of Kielmeyer and his pupil, the physiologist Johann Heinrich Ferdinand von Autenrieth.

Kielmeyer's lectures in Stuttgart and Tübingen 'circulated in countless copies throughout Germany'.[14] Many of these are still to be found in Kielmeyer's papers, which are preserved in Stuttgart and Tübingen as well as in other libraries. These originate from all the periods in which Kielmeyer held lectures, and cover all fields of his teaching: general zoology (or 'the physics of organic bodies'), chemistry, botany, physiology of plants, comparative anatomy and materia medica. The wide distribution of these lecture notes can be explained by the fact that, over time, they were replicated by copyists and then sold. One such manuscript was acquired by Arthur Schopenhauer, who then extracted excerpts from it for his own studies.[15]

In this way, the contents of Kielmeyer's lectures reached a later generation of students who would not have been able to attend themselves. As the zoologist Rudolph Wagner reported, he, along with the chemist Christian Friedrich Schönbein and many others, first became acquainted with Kielmeyer's ideas in the 1820s.[16] We can guess that the copy on which the unauthorized publication of Kielmeyer's lecture on *Allgemeine Zoologie oder Physik der organischen Körper*[17] was based also found its way by these means to the prosector Gustav Wilhelm Münter at Halle. Münter published the manuscript in book form during Kielmeyer's lifetime without revealing the true author of the text.[18] In Kielmeyer's unpublished correspondence, there is clear evidence that, not only the plagiarist Münter, but also several more renowned colleagues used Kielmeyer's lectures, at least partially, in their own textbooks.

Kielmeyer's Reports

The fact that Kielmeyer published little during his active period as professor and scientist did not prevent him from climbing to great heights on the career ladder. At the age of fifty-one, he became the director of scientific collections (library, botanical garden, and natural history cabinet) and was appointed privy counsellor in Stuttgart. He entered into this newly created position in June 1817. This post – a sort of scientific secretary of state – included many additional administrative duties that Kielmeyer had to execute. At the same time, it enabled him to advise on many medical, scientific, technological and political matters throughout Württemberg.

Having already compiled such reports for the faculty and the ministry during his time as professor at Tübingen, he was now regularly questioned on diverse topics in all the fields of his expertise, such as on the founding of the Polytechnic School (later the University of Stuttgart) in the year 1829.[19] As with his lectures, the majority of these reports never appeared in print, but several did form the scientific basis for political decision-making. When not being exploited for public policy, Kielmeyer's expert documents were merely put on file, the one exception being a published assessment of lightning and hail conductors made from straw ropes.[20]

Moreover, as with his lecture notes, a considerable number of Kielmeyer's reports are to be found in handwritten form. They make up a further large, and up to now practically unknown, part of his scientific oeuvre. Six of these were published posthumously (four forensic, one psychiatric and one technological report). More than twenty further assessments, covering a wide range of topics, still await publication. They reveal, for example, the extent to which Kielmeyer expanded the scientific collections of the library, the botanical garden and the natural history cabinet with new acquisitions. Their most important role, however, could be to give an insight into his day-to-day work as a scientific administrator and political advisor in Württemberg, for we know next to nothing about his achievements in this influential post which he held for twenty-two years before his retirement in 1839.

Kielmeyer's Fame

In the history of medicine and natural science, the fame of a researcher is closely linked to his publications.[21] In the eighteenth century, the scientific quality of an academic was judged by contemporaries according to two criteria, and success in this respect led to an increased chance of being appointed to a university. One was the compilation of the fundamental subject matter that encompasses one field of study into a textbook. The other was the publication of original papers in increasingly disciplinary-oriented natural science journals – and this latter criterion went on to become the norm in determining a scientist's significance.

In Kielmeyer's case, things were somewhat more complicated. With just one thesis, his medical dissertation from 1786, he fulfilled the minimum necessary for his appointment to the Hohe Karlsschule. But his attitude when it came to further publishing was unusual for a professor on the career ladder. Kielmeyer would later refuse many honourable calls to chairs. Only his short lecture on organic forces with its essay-like character brought him, within just a few years, the literary fame which others only achieved after the publication of numerous papers and books. In Karl Heinrich Ludwig Pölitz' anonymous survey of the most important contemporary philosophers in 1800, *Die Philosophie unseres Zeitalters in der Kinderkappe*, there is an entry on 'Kielmeyer, (professor in Tübingen)':

> In a time when philosophy first began to illuminate the great doctrines of organization, this man stepped up with one essay (more a talk) that caused a sensation, and which left us eager for his promised (but yet unpublished) work:

History and Theory of the Development of Organisations [*Geschichte und Theorie der Entwickelung der Organisationen*]. These few sheets are strewn with stimulating thoughts.[22]

Despite Kielmeyer's announcement that a larger work expounding his thoughts was pending (p. 31), and although several contemporaries had publicly prompted him to publish[23], the promised text was not published in his lifetime (it was first published by Holler in *GS*, 102–94). As a consequence, at the turn of the nineteenth century, Kielmeyer's work became surrounded by an aura of secrecy. And some of his most renowned contemporaries, like Goethe, Schelling and the Humboldt brothers, testified of their high esteem for him.

Goethe visited Kielmeyer in Tübingen in 1797 and spoke with him about 'various aspects of anatomy and physiology of organic nature'. There Kielmeyer advanced his ideas of 'how he could connect the laws of organic nature with general physical laws'.[24] As early as 1798, Schelling proclaimed in *On the World-Soul* that Kielmeyer's lecture would later be recognized as the start of 'a wholly new epoch in natural history'.[25] Alexander von Humboldt dedicated his *Beobachtungen aus der Zoologie und vergleichenden Anatomie* to the 'great physiologist Kielmeyer' as a sign of 'deep admiration and great respect'.[26] In 1810, his brother Wilhelm von Humboldt, in reorganizing the university system in Prussia, tried to call him to the newly founded University of Berlin, for 'nearly all the good new physiological ideas originate from Kielmeyer in Tübingen'.[27] Other notable personages of the nineteenth century, among them Cuvier, the physiologist Johannes Müller, Schopenhauer and the geologist Henrich Steffens, also stressed the importance of Kielmeyer's influence on their own work.

Kielmeyer was seen by his peers as the 'ideal' (*Musterbild*) German university teacher.[28] Even despite his reticence to publish, he was offered posts at the most important universities of the Enlightenment era: Halle (1803/04)[29] and Göttingen (1804) as well as the newly founded University of Berlin (1810). As esteemed as these invitations were, he refused them all and remained loyal to his native Württemberg.

Kielmeyer was awarded practically all the honours of the time which a scientist could receive. He was a fellow of twenty-two scientific academies and research communities, including ten honorary memberships.[30] The academies of science in Munich (1808) and Berlin (1812) elected him as their member as did the *Kaiserliche Leopoldinisch-Carolinische Deutsche Akademie der Naturforscher Leopoldina* (1816) and the *Académie de médecine* in Paris (1836). After Kielmeyer's death on 24 September 1844, both the Munich Academy and the Leopoldina produced extensive obituaries.[31] However, it is surprising that Kielmeyer never became a fellow of the Göttingen Academy of Sciences nor of the French *Académie des Sciences* – maybe this was because of his lack of publications or possibly he was overlooked. Academics were usually chosen in their later years to become members of learned societies and, coupled with his relatively early move, his position as privy counsellor may have been perceived as of a more administrative nature.

In addition, Kielmeyer received highest civil honours within his native Württemberg. In 1808, King Frederick I of Württemberg awarded him a knighthood together with a medal. In September 1834, Kielmeyer presided over the twelfth meeting of the *Versammlung deutscher Naturforscher und Ärzte* in Stuttgart, where he gave his last ever public talk, a lecture concerning the growth of plants, in which he picked up on the theme of his famous 1793 Speech. In 1835, he was given honorary citizenship by the city of Stuttgart and, as early as 1826, a genus of Brazilian flowering plant was named *Kielmeyera* by the botanist Karl Friedrich Philipp von Martius.

Written in an unknown hand on Pfaff's notes taken at Kielmeyer's zoology lectures, a distich can be found which was later used in Kielmeyer's obituary: 'What Aristotle is to the Greeks, Harvey to the Britons, Kielmeyer will be for the Germans' ('Graiis Aristoteles decus Harvaeusque Britannis. Teutonum populus Kielmeyerus erit').[32] In 1938, when the German physicians and naturalists once again convened in Stuttgart, a plaque was unveiled paying homage to the 'herald of evolutionary thinking' (*Künder des Entwicklungsgedankens*)[33].

Kielmeyer's Fate

Kielmeyer's fame stands in inverse proportion to the number of his scientific publications. He was infamous precisely insofar as he published so little. Schopenhauer, for example, writes, 'Kielmeyer has, as is well-known, published nothing'.[34] The expression 'publish or perish' did certainly not apply here. For several decades his renown was founded on the reputation of his lectures and talks, the Göttingen physiologist Rudolph Wagner describing him as 'one of the quietly famous, who, without having written anything significant, gives mighty impulses into their surroundings which only become effective when taken up by others who are of more active and communicative natures'.[35] However, Wagner also drew attention to the 'destiny of ageing scholars', who, when no longer able to follow the fastmoving developments of the sciences, 'soon go unrecognised, and without renown are cast aside'.[36]

The predominant picture that posterity has of Kielmeyer's achievements is in great part influenced by his reluctance to publish his substantial knowledge. Only one seminal publication justified his fame and exerted a strong influence on medicine, natural sciences and philosophy around 1800. Further claims found in his smaller publications and in the dozens of doctoral dissertations of his students were, in comparison, less influential. Many of his students, some of whose doctoral theses he supervised as *praeses*, used ideas from his lectures in their dissertations and named their source explicitly.[37] Until now, however, secondary literature on Kielmeyer has only taken sporadic notice of these dissertations. One example is Johann Christian Salomo Tritschler's dissertation from 1807 which received particular attention because it contained a hypothesis of organic evolution (*hypothesi de evolutione organica*). Tritschler saw the series of animals as 'continuous and directly ascending, with each member distinguished by its grade of evolution and heterogeneity of its organs'.[38]

Even today, preoccupation with Kielmeyer's work is based on just a very small number of texts. The only edition of his works (*GS*) prepared by Fritz-Heinz Holler, contained a very limited selection from the few printed sources and some hitherto unpublished material. Without any textual criticism or commentary, this edition already fell short of modern editorial requirements at the time of its publication. Similarly, Coleman complained that his biographer 'must deal with Kielmeyer's scientific views only through quite inadequate published material', and he added that there is 'no census or any certain ed. of lecture notes kept by Kielmeyer's students'.[39] To date only Max Rauther, who compared several versions of lecture notes on general zoology or the physics of organic bodies[40], and Reinhard Löw, who edited parts of a chemistry lecture[41], have made systematic use of these lecture notes.

For a deeper understanding of Kielmeyer's actual achievements, it is essential to extend the current source-base and include further texts, both the printed and the unpublished papers. The greater part of Kielmeyer's oeuvre still remains in handwritten form or is to be found in the publications of his students. Only when these have been taken into account will it be possible to determine more precisely his place in medicine, philosophy and the sciences at the turn of the nineteenth century.

Historiographical Perspectives

Hence, the very different interpretive approaches to Kielmeyer and his thought that are to be found in the scholarship are only partly satisfactory. It is surprising and somewhat contradictory that Kielmeyer, who was called the 'father of philosophy of nature' by Cuvier,[42] has also been credited in recent years as the first person to envisage the possibility of 'biology' as a general, unifying branch of science engaged with laws that govern the organic world:

> Kielmeyer used a similar framework to explain self-organization not in the individual body but for organic nature in its entirety. In this sense, though not as explicitly as Treviranus a few years later, Kielmeyer's *Rede* addresses for the first time the possibility of biology as a general, unified field – one concentrated with the laws that regulate the organization of all living nature.[43]

In Kielmeyer's time, 'physiology' was still the proper term for a science of the laws of life. When it was put into question, it was by Erasmus Darwin's zoonomy ('zoonomia'), rather than by the newly coined term 'biology'. Certainly, Kielmeyer would have been fully aware of the different approaches to a new science of life in the decades around 1800, but, as far as we know, he never used the word 'biology' or made a contribution to this discussion. It is true that his concept of a harmony (and hierarchical order) of organic forces should be seen as a further development of the notion of a life force (*Lebenskraft*) so in fashion in the late eighteenth century. It opened up the possibility of using such a concept as the core and basis of a new life science. Thus, Kielmeyer's concept of organic forces is ultimately quite easy to connect to a new science of life or 'biology' – as Schelling almost proclaims, when he speaks of the 'epoch of a wholly new natural history'.[44]

With Kielmeyer's promotion to what may be called the 'father of biology', another concept is overlooked which had long been predominant in the scholarship. Notwithstanding that he had, in William Harvey and Aristotle, two prominent forerunners when it came to the concept of recapitulation, it has been common for most scholars to state, ever since the nineteenth century, that it was Kielmeyer who first introduced recapitulation into biology – that is, the parallel between stages of development of an embryo and successive adult stages in evolution. This notion became especially popular after the triumphal advance of Charles Darwin's evolutionary theory and the formulation of the 'biogenetic law' by Ernst Haeckel in 1866, which states that ontogeny recapitulates phylogeny. Throughout the twentieth century, Kielmeyer was mentioned in almost all biological works (including histories of biology) as the most important forerunner of Haeckel. This trend begins with Thomas Hunt Morgan who states that 'the first definitive reference to the recapitulation view . . . I have been able to find is that of Kielmeyer in 1793'.[45] E. S. Russell put it more cautiously, 'The law of parallelism seems to have been expressed first by Kielmeyer (1793)'[46], while Steven Jay Gould stressed that 'most prefer Kielmeyer 1793, as did Meckel (1821) in the first attempt I know to establish a chronological list of recapitulation's supporters'.[47] Even if Kielmeyer was not included in the classic anthology *Forerunners of Darwin*[48], he was still viewed during the twentieth century as one of the more important 'precursors of evolutionary thinking'.[49] It is remarkable, therefore, that, in the last few years, Kielmeyer's significance has no longer been primarily based on his role in the history of evolutionary biology.

Beginning with Timothy Lenoir's *The Strategy of Life*[50], a new appreciation of Kielmeyer's significance has led to a re-evaluation of his position in the history of medicine and science at the turn of the nineteenth century. Lenoir called a group of Blumenbach's disciples – Alexander von Humboldt, Heinrich Friedrich Link, Gottfried Reinhold Treviranus as well as Kielmeyer himself – 'the Göttingen school of biology'.[51] All had attended lectures by Blumenbach. In their writings, Lenoir identified a common frame of thinking – they all shared a 'teleomechanistic' worldview, based on Kant's critical philosophy.

The question as to what extent, or even if, such a 'Göttingen School of Biology' existed and, if so, Kielmeyer's connection to that school remains a matter of controversy.[52] There is no indication that contemporaries were aware of a 'Göttingen school', nor has the historiography on Blumenbach or Kielmeyer made such a claim. This does not mean that Lenoir's conclusion is false. Rather, his arguments need to be verified and, where necessary, refined. In the case of Kielmeyer, his limited personal contact with Blumenbach does not really support the thesis that he was one of his particularly close students. A recently discovered letter from Blumenbach to Kielmeyer (10 October 1808)[53] shows their mutual appreciation and respect for each other, but does not refer to any scientific matters. Furthermore, each possessed quite different philosophical backgrounds, as Frank

Dougherty[54] has explained in detail. Dougherty also stresses that Kielmeyer did not appropriate Blumenbach's central notion of a formative power (*Bildungstrieb*) for his concept of organic forces. For these reasons, I agree with Zammito's recent conclusion that Kielmeyer 'started from a different perspective' than Blumenbach and 'he advanced it in directions that were clearly his own'.[55]

This different perspective can be found in the work of the theologian Johann Gottfried Herder, who was also Kielmeyer's 'favourite poet' (*Lieblingsdichter*).[56] A close reading of Herder's historical-philosophical *Ideas for a Philosophy of the History of Mankind* demonstrates that Kielmeyer provided the detailed physiological arguments for Herder's reflections on the relation of forces in nature. Herder called for 'a philosophical anatomist' (*Zergliederer*) to undertake a 'comparative physiology' of several 'distinctive and definitive forces' in relation 'to the whole organization of the creature'.[57] This task was immediately taken up by the young Kielmeyer. In 1799, Kielmeyer's contemporary Schelling already identified the decisive influence of Herder – and not Blumenbach – in Kielmeyer's speech on organic forces; he even states that Kielmeyer took his 'main idea' from Herder.[58] Accordingly, this concept has subsequently come to be called the 'Herder-Kielmeyer tripartite division of life functions into reproduction, irritability and sensibility'[59], and it is therefore apt that several historians have stressed Kielmeyer's debt to Herder, particularly in his 1793 Speech. 'Here was Kielmeyer's program', William Coleman[60] has stated, and Zammito writes, 'Herder was a decisive inspiration for Kielmeyer's project'.[61]

Lenoir dedicates a whole section to 'Carl Friedrich Kielmeyer and the Physik des Thierreichs' in *The Strategy of Life*,[62] and, of course, over the last few years whole chapters of books have been devoted to him. For example, Robert J. Richards includes a chapter on 'Kielmeyer and the Organic Powers of Nature' in *The Romantic Conception of Life*.[63] Kielmeyer has come to earn a place beside such contemporaries as Blumenbach, Johann Christian Reil, Goethe and Schelling. Over the last couple of years Kielmeyer has been put at the center of any account of the history of biology around 1800. Hence, the most recent readings of Kielmeyer in Gambarotto and Zammito describe the shift in focus from natural history and physiology to 'biology' at the turn of the nineteenth century in a manner that puts Kielmeyer's conception of organic forces at the centre.[64] Furthermore, Gambarotto has also identified Kielmeyer's lecture as 'the program for a general biology', seeing in it a new and unifying field of science.[65]

Another topic in the history of science that has been popular in recent years is Kielmeyer's significant contribution to the formation of comparative method. Johannes Müller already claimed in Kielmeyer's lifetime that 'Kielmeyer was the first to conceive comparative anatomy from its innermost aspect'; he had 'called it to life' and given it 'its intellectual orientation'. He explained that it 'follows nature in the procreation of the living process of production'.[66] Kielmeyer's comparative approach went far beyond the study of anatomy. He compared the hierarchy of forces as well as the distribution of matter. The laws he found in the realm of the organic he closely linked to the laws ruling the inorganic, especially to electricity and magnetism. Furthermore, he correlated the development of life to the history of the earth in his search to discover the one fundamental force that causes the evolution of the world (p. 43) – a goal unattainable in his time.

Shortly after his death, he was acclaimed as the father of 'comparative science': 'Kielmeyer possessed the idea of the unity of nature as well as the empirical evidence of the relation of the different forms and processes of nature'.[67] In particular, his perseverance in using and executing a 'comparative method'[68] was central to the rediscovery of him as a key figure in the development of scientific methodology. Kielmeyer and Cuvier are now viewed as equally responsible for comparative method in anatomy[69], and the 'application of the comparison in natural science' is claimed to have reached with Kielmeyer 'maybe its highest point'.[70] And what is so praised in this comparative method is that he used many analogies, as did the philosophers of nature who followed him, but yet he did not neglect the empirical basis of his research:

> In his study of natural history, Kielmeyer applied both detailed comparison and systematic ordering of objects, and also reasoning by analogy, to explain the interrelationship of nature in developmental history through hypothetical forces.[71]

Conclusion

As we have seen, Kielmeyer's intellectual biography is influenced by many different sources. He certainly had a mind of his own, for his ideas cannot be reduced to a product of his education. It is astounding that he taught himself precisely the two sciences that became indispensable for his later work, i.e. chemistry and physiology. We have to look well beyond the so-called 'Göttingen School' to fully recognize Kielmeyer's genius. Many aspects of his life and work still remain to be discovered.

Precisely because of his reluctance to publish his lectures or other scientific papers, Kielmeyer himself contributed much to the ambiguity of his place in history. A scholar who published 'virtually nothing'[72] can easily be used to support a variety of interpretations. This vagueness of his published work, particularly the short 1793 Speech on organic forces, leads to all kinds of interpretations, and. for most commentators the 1793 Speech was – and still is – the starting point for some 'new epoch in natural history', for a general biology, for evolutionary thinking (including the idea of recapitulation), for a comparative approach in science, for a doctrine of the hierarchy of forces, and, last but not least, for many kinds of speculative thinking by the Romantic *Naturphilosophen*. This makes his 1793 Speech so appealing for a rethinking of the relation of philosophy and the sciences at the turn of the nineteenth century.

Kielmeyer's oeuvre, it must again be emphasized, extends much beyond this sensational speech. Most of his students' publications, starting with their doctoral theses, were inspired by Kielmeyer's thinking, as they learnt of it during his lectures. And, in many cases, he is named explicitly in their papers. To obtain an overview of Kielmeyer's work, the contents of these publications need to be taken into consideration. Moreover, the majority of these extant lecture notes deserve a much more thorough study, and a critical edition of a selection of them would be invaluable in reconstructing Kielmeyer's teachings. Finally, his unpublished assessments and reports which cover a wide range of topics must also be seen as a vital component of his scientific legacy.

In recent years, the re-evaluation of Kielmeyer's works and of their impact on the philosophical and scientific discussions taking place around 1800 has paid scant attention to his unpublished papers. Without exploring these sources to reassess his scientific achievements a full understanding of their influence will be impossible. One result of further research may be to show that not all representations of Kielmeyer that are currently associated with him hold true in the end. Would the man who rejected the claim of being the 'Father of *Naturphilosophie*' be proud to now be labelled the 'Inventor of the Biogenetic Law' or the 'Father of Biology'? Probably not. Perhaps the 'real' Kielmeyer was a quite different scientific persona from the image most of us have of him.

Notes

1 S. Schmitt (ed.), *Les forces vitales et leur distribution dans la nature. Un essai de 'systématique physiologique'* (Turnhout: Brepols, 2006), 107-28.
2 F. Buttersack, 'Karl Friedrich Kielmeyer (1765-1844). Ein vergessenes Genie', in *Sudhoffs Archiv für Geschichte der Medizin*, 23 (1930), 236-46.
3 A. K. Mayer, 'Eine Reliquie von C. Frd. Kielmeyer', in *Archiv der Heilkunde* 5 (1864), 354-67; 'Nachtrag', in *Archiv der Heilkunde* 6 (1865), 474-5.
4 J. F. Blumenbach,, *Handbuch der Naturgeschichte* (Göttingen: Dieterich, 1779/80).
5 See K. T. Kanz, 'Die Naturgeschichte (Botanik, Zoologie, Mineralogie) an der Hohen Karlsschule in Stuttgart (1772-1794)', in *Jahreshefte der Gesellschaft für Naturkunde in Württemberg* 148 (1993), 5-23.
6 K. T. Kanz, 'Carl Friedrich Kielmeyer, Lichtenberg und Göttingen 1786-1796', in *Lichtenberg-Jahrbuch* (1989): 142.
7 Ibid., 145.
8 F. W. P. Dougherty, 'Über den Einfluß Johann Friedrich Blumenbachs auf Kielmeyers feierliche Rede von 1793. Mit einem Anhang über Kielmeyers Göttinger Lektüre', in K. T. Kanz (ed.), *Philosophie des Organischen in der Goethezeit* (Stuttgart: Steiner, 1994), 67-80.
9 Kanz, 'Die Naturgeschichte', 16.
10 Translated in the next chapter.
11 C. H. Pfaff, *Lebenserinnerungen* (Kiel: Schwers, 1854), 64.
12 See K. T. Kanz, *Kielmeyer-Bibliographie. Verzeichnis der Literatur von und über den Naturforscher Carl Friedrich Kielmeyer (1765-1844)* (Stuttgart: Verlag für Geschichte der Naturwissenschaften und der Technik, 1991), 32-40.
13 K. T. Kanz, '"Halle zum vollkommnen Muster einer Academie zu organisieren." Johann Christian Reils Berufungskorrespondenz mit J. H. F. Autenrieth und C. F. Kielmeyer in Tübingen', in F. Steger (ed.), *Johann Christian Reil. Universalmediziner, Stadtphysikus, Wegbereiter von Psychiatrie und Neurologie* (Gießen: Psychosozial-Verlag, 2014).

14 R. Wagner, *Samuel Thomas Soemmerrings Leben und Verkehr mit seinen Zeitgenossen*, vol. 2 (Leipzig: Voss, 1844), 164.
15 A. Hübscher, 'Kleine Nachlese', in *Schopenhauer-Jahrbuch* (1983): 154–64.
16 Wagner, *Samuel Thomas Soemmerrings Leben*, vol. 2, 164.
17 G. W. Münter, *Allgemeine Zoologie oder Physik der organischen Körper* (Halle: Schwetschke, 1840).
18 See I. Jahn, 'War Gustav Wilhelm Münter (1804–1870) ein "Plagiator" Kielmeyers?', in K. T. Kanz (ed.), *Philosophie des Organischen in der Goethezeit* (Stuttgart: Steiner, 1994); and K. T. Kanz, '"Wie einzelne Menschen, so haben auch einzelne Entdeckungen ihr besonderes Glück." Gustav Wilhelm Münter (1804–1870) und seine Entdeckung der "Physik der organischen Körper" oder der wissenschaftliche Größenwahn eines anatomischen Kustos', in J. Schulz (ed.), *Fokus Biologiegeschichte. Zum 80. Geburtstag der Biologiehistorikerin Ilse Jahn* (Berlin: Akadras, 2002).
19 J. H. Voigt, 'Lehre zwischen Politik und Wirtschaft 1829–1864. Von der Real- und Gewerbeschule zur Polytechnischen Schule', in *Festschrift zum 150jährigen Bestehen der Universität Stuttgart. Beiträge zur Geschichte der Universität* (Stuttgart: Deutsche Verlags-Anstalt, 1979), 29.
20 C. F. Kielmeyer, 'Gutachten des Herrn Staatsraths von Kielmeyer über die Lapostolle'sche Schrift, betreffend Blitz- und Hagelableiter aus Strohseilen', *Correspondenzblatt des Würtembergischen Landwirthschaftlichen Vereins* 7 (1825): 268–76.
21 F. Stuhlhofer, *Lohn und Strafe in der Wissenschaft. Naturforscher im Urteil der Geschichte* (Vienna: Böhlau, 1987).
22 [K. H. L. Pölitz], *Die Philosophie unseres Zeitalters in der Kinderkappe; von einem Manne, der auch lange in dieser Kappe gelaufen ist* (Pirna: Arnold & Pinther, 1800), 351.
23 K. T. Kanz, '"Die Besten der Gelehrtenwelt beklagen Ihr Stillschweigen": Schellings Einladung an Carl Friedrich Kielmeyer zur Mitarbeit an den *Jahrbüchern der Medicin als Wissenschaft*', in *Schelling-Studien: Internationale Zeitschrift zur klassischen deutschen Philosophie* 5 (2017): 189–90.
24 J. W. Goethe, *Tagebücher. Bd. II,1. 1790–1800. Text*, ed. E. Zehm (Stuttgart: Metzler, 2000), 177.
25 F. W. J. Schelling, *Von der Weltseele, eine Hypothese der höhern Physik zur Erklärung des allgemeinen Organismus* (Hamburg: Perthes, 1798), 298.
26 A. Humboldt and A. Bonpland, *Beobachtungen aus der Zoologie und vergleichenden Anatomie* (Tübingen, Paris: Schoell, 1809), [v].
27 R. Köpke, *Die Gründung der Königlichen Friedrich-Wilhelms-Universität zu Berlin* (Berlin: Schade, 1860), 201.
28 C. F. P. Martius, *Akademische Denkreden* (Leipzig: Fleischer, 1866), 183.
29 See Kanz, 'Halle zum vollkommnen Muster einer Academie zu organisieren'.
30 G. F. Jaeger, 'Ehrengedächtniss des königl. württembergischen Staatsraths von Kielmeyer', in *Novorum Actorum Academiae Caesareae Leopoldino-Carolinae Naturae Curiosorum* 21.2 (1845), xxxv–vi.
31 See Jaeger, 'Ehrengedächtniss', and Martius, *Denkreden*.
32 Jaeger, 'Ehrengedächtniss', xlii.
33 M. Rauther, 'Carl Friedrich Kielmeyer zu Ehren', in *Jahreshefte des Vereins für vaterländische Naturkunde in Württemberg* 94 (1938), xxv.
34 A. Schopenhauer, *Der handschriftliche Nachlaß in fünf Bänden*, vol. 5, ed. Arthur Hübscher (Munich: Deutscher Taschenbuch Verlag 1985), 265.

35 Wagner, *Samuel Thomas Soemmerrings Leben*, vol. 2, 164.
36 Ibid.
37 E.g. in K. T. Kanz, 'Die Besten der Gelehrtenwelt', 190–1.
38 Buttersack, 'Kielmeyer', 237; English trans.: R. J. Richards, *The Romantic Conception of Life. Science and Philosophy in the Age of Goethe* (Chicago: University of Chicago Press, 2002), 246.
39 W. Coleman, 'Kielmeyer, Carl Friedrich', in C. C. Gillispie 1973 (ed.), *Dictionary of Scientific Biography*, vol. 7, 367–8.
40 M. Rauther, 'Ungenutzte Quellen zur Kenntnis K. Fr. Kielmeyers', in *Besondere Beilage des Staatsanzeigers für Württemberg* 6 (14 May 1921), 113–22.
41 R. Löw, *Pflanzenchemie zwischen Lavoisier und Liebig* (Munich: Donau-Verlag, 1977), Anhang III-IV: 10–65.
42 G. Cuvier, 'Dernière leçon.' *Le Temps* 5 June 1832, 14454.
43 A. Gambarotto, *Vital Forces, Teleology and Organization. Philosophy of Nature and the Rise of Biology in Germany* (Dordrecht: Springer, 2018), 46.
44 Schelling, *Von der Weltseele*, 298. See N. Jardine, 'The significance of Schelling's "Epoch of a wholly new natural history". An essay on the realization of questions', in R. S. Woolhouse (ed.), *Metaphysics and Philosophy of Science in the Seventeenth and Eighteenth Centuries* (Dordrecht: Reidel, 1988).
45 T. H. Morgan, *Evolution and adaptation* (New York: Macmillan, 1903), 57–8.
46 E. S. Russell, *Form and Function. A Contribution to the History of Animal Morphology* (Chicago: University of Chicago Press, 1982), 90.
47 S. J. Gould, *Ontogeny and Phylogeny* (Cambridge, MA: Belknap Press, 1977), 35.
48 B. Glass, O. Temkin and W. L. Straus (eds), *Forerunners of Darwin 1745–1859* (Baltimore: Johns Hopkins Press, 1959).
49 K. T. Kanz, 'Carl Friedrich Kielmeyer (1765–1844) – Wegbereiter des Entwicklungsgedankens', in H. Albrecht (ed.), *Schwäbische Forscher und Gelehrte. Lebensbilder aus sechs Jahrhunderten* (Stuttgart: DRW-Verlag, 1992).
50 T. Lenoir, *The Strategy of Life: Teleology and Mechanics in Nineteenth-Century German Biology* (Dordrecht: Reidel, 1982).
51 Ibid., 115.
52 See Gambarotto, *Vital Forces*, 33–5; and J. Zammito, *The Gestation of German Biology. Philosophy and Physiology from Stahl to Schelling* (Chicago: University of Chicago Press, 2018), 254.
53 Kanz, 'Die Besten der Gelehrtenwelt', 187.
54 Dougherty, 'Über den Einfluß Johann Friedrich Blumenbachs'.
55 Zammito, *Gestation*, 255.
56 Pfaff, *Lebenserinnerungen*, 42.
57 J. G. Herder, *Ideen zur Philosophie der Geschichte der Menschheit*, ed. W. Pross (Munich: Hanser, 2002), 221; English trans.: Zammito, *Gestation*, 250.
58 F. W. J. Schelling, *Erster Entwurf eines Systems der Naturphilosophie zum Behuf seiner Vorlesungen* (Jena: Gabler, 1799), 221. See Zammito, *Gestation*, 250.
59 C. Siegel, *Geschichte der deutschen Naturphilosophie* (Leipzig: Akademische Verlagsgesellschaft, 1913), 238.
60 Coleman, 'Kielmeyer', 367.
61 Zammito, *Gestation*, 250.
62 Lenoir, *Strategy of Life*, 37–63.
63 Richards, *Romantic Conception of Life*, 237–51.
64 Zammito, *Gestation*, 245–85.

65 Gambarotto, *Vital Forces*, 42–6.
66 J. Müller, *Zur vergleichenden Physiologie des Gesichtssinnes des Menschen und der Thiere* (Leipzig: Cnobloch, 1826), 29.
67 F. Fischer, *Die Metaphysik, von empirischem Standpunkte aus dargestellt* (Basel: Schweighäuser, 1847), 30–1.
68 I. Schumacher, 'Karl Friedrich Kielmeyer, ein Wegbereiter neuer Ideen. Der Einfluß seiner Methode des Vergleichens auf die Biologie der Zeit', in *Medizinhistorisches Journal* 14 (1979), 81–99.
69 M. Eggers, *Vergleichendes Erkennen. Zur Wissenschaftsgeschichte und Epistemologie des Vergleichs und zur Genealogie der Komparatistik* (Heidelberg: Winter, 2006), 231–70.
70 Ibid., 19.
71 Ibid.
72 T. Lenoir, 'Kant, Blumenbach, and Vital Materialism in German Biology', *Isis* 71 (1980), 108.

Part I

KIELMEYER'S 1793 SPEECH

Chapter 3

ON THE RELATIONS BETWEEN ORGANIC FORCES
IN THE SERIES OF DIFFERENT ORGANISATIONS,
AND ON THE LAWS AND CONSEQUENCES
OF THESE RELATIONS[1]

C. F. Kielmeyer

[1] Esteemed audience.
A person's date of birth determines so much of what we attribute to their nature and self-acquired properties that people's custom of periodically remembering theirs with celebration is more than a little justified.

If it so happened, in the course of things, that one person had at his command the circumstances and changes of fortune of others, then the interest in remembering this date increases for those others whose circumstances are indirectly determined by him.

Today we have the opportunity to observe one such moment of time come around again for the Prince[2] to whom this academy owes its existence and its development, and to whom every member of the academy more or less owes the situation in which they now find themselves. [2] It is so natural to celebrate this day by joining our remembrance of it with feelings of joy and gratitude and by longing for the return of these feelings when the date comes round again that each of us, if no one else, will be disposed to reflect in this way without any prompting. An invitation to do so, issued by me, is thus unnecessary – all the more unnecessary, the more certain it is that any emotion, once spoken, thereby becomes an empty one, and there is nothing company is less able to tolerate than precisely those inner feelings that rush upon us. It seems more fitting for today's celebration to lay out a series of ideas, which, even if they are not at first glance related to these feelings, nevertheless serve to support them, due to their generally enduring nature, and so can provide some permanence to the usually fleeting and transitory existence [of such feelings]. And because our intellectual interest increases roughly in proportion to the greatness of the object that our ideas encompass, it seems most fitting to propose an object that, by its own greatness, imparts to them the interest they themselves lack, and thereby works all the more surely towards the attainment of the above goal.

* * *

[3] If, by the powers of our minds, we separate the phenomena [*Erscheinungen*[3]] of nature – for us connected in a system by space and time – from their connection, then surely those phenomena that we isolate and subsume under the name 'animate nature'[4] – I mean the organisations[5] of our earth – are those most able to fill us with feelings of nature's greatness among [the phenomena] with which we are closely acquainted. To be sure, no masses, volumes, or distances found here are like those of the skies, by which nature convinces us of its greatness. However, if, when judging the greatness of an object, we can deign to give voice and listen with a little patience to the multiplicity, diversity and harmony of effects in a small space and short periods of time, then there are things of another kind that speak to us no less forcefully.

Here, one could say that in the small space which the surface of the earth and its relation to the sun allows, seven million different bodily forms are gathered (according to one very moderate estimate); each multiplied (according to the lowest possible estimate) into 10,000 individuals, and every individual composed of a slew of different organs, [4] whose number, Lyonet[6] has shown, in smaller and simpler individuals often reaches the thousands and up to ten thousand. And if what is shown to us in this small space isn't enough, then one should also consider how all these things occupy time: in the changes that it experiences at every moment, each organ is so adapted to the changes of all other organs, and so united in a system of simultaneous and consecutive changes, that – according to our manner of speaking – each is reciprocally cause and effect of the other.[7] Each of the living individuals, thus animated by their organs, endures for a greater or lesser stretch of time, and, at each point in this course of time, the system of effects that we call its life, and the system of organs that constitute its organism, change themselves, one emerging from the other as its cause. In every individual, infancy, youth, old age, and death lead on one from another; and, in each of these states, those individual's effects stand connected anew to the effects of other individuals of the same species in a greater system of effects. The infancy of one depends on the maturity of the other, and the youth of one on [5] the youth of the other. This connection is so tight that we should believe, according to our manner of speaking and representing, that nature had interwoven the nerves of an individual with those of the others into a web, and the impressions of one would be felt in the sensorium of the other. If we wanted to take the trouble to look merely at the history of the human species, we would soon see that this greater system of effects that one may call the life of the species gradually progresses along a path of development over greater periods of time; we would also see that, although history permits us to observe only a small element of this path, the infancy of our species is only now turning towards the dawn of a joyful youth.

Finally, the effects of the individuals of a species are linked together in a system of effects with the effects (to which they are so often opposed) of individuals of other species to form the life of the great machine of the organic world. This machine also appears to be progressing along a path of development that we may best represent with the image of a parabola that never closes in on itself. To all the preceding, one should still add the following little detail: suppose that nature [6] had no intent in artificially placing phenomena one after another and next to each other in time, that those effects and consequences were not ends [*Zwecke*] that [nature] wanted to achieve, and suppose it an empty reverie to want to assume higher ends unknown to us – [even then] we would still nevertheless have to admit that in most cases this chain of cause and effect looks like a chain of means and end [*Absicht*] to us, and we even find it conducive to our reason to accept such a chain; and we must thereby, at least in the end, admit that nature is no less able here than in the skies of convincing us of the truth of that [the greatness of nature] from which I began.

* * *

In this overview of the phenomena of living nature, the greatest effect, the one which activates our attention most, must be the fact that, irrespective of the many conflicting forces in it, so far as we can tell, as a whole nature always remains the same, and nature's quiet course goes on unhindered. The question of the causes and the forces by which this effect is obtained [7] is the most pressing for us. Since we are justified in seeking these causes in those effects [that are] notable in the individual and in their relation [*Verhältnis*[8]] to one another, the following questions now arise:

First, which forces are united in the greatest number of individuals?

Then, what are the proportions [*Verhältnisse*] of these forces to each other for different species of organisations, and by what laws do these proportions modify themselves in the series of different organisations?

And finally, how are the effects and consequences previously alluded to – namely the development and continued existence of this organic world, and the species that compose it – grounded in [the forces] as their cause?

I have taken little trouble in choosing as the object of this discussion to answer these questions and, above all, the second question in a few fragments, and to present some of the results revealed to me in the search for pertinent laws. The less trouble I have taken in so doing, the more the simplicity of the means and causes that nature sets in motion in order to preserve its effects [8] is capable of heightening our conception of its greatness, and the more the greatness and difficulty of [its] objects excuses the errors in the attempt.*

---

* *Remark*. The whole essay, as it appears here, would surely have remained unprinted if the occasion for which it was made and the request of a few people had not changed my mind on the matter. Because mere fragments are presented here, and thus the assessment itself cannot be complete, I must ask that judgement on many of the ideas propounded here (if the paper should reach people who could and must be accepted as judges) be postponed. Because I seek to explain, expand, and amend these and other ideas in a history and theory of the development of organisations more closely, if time and health permit, I hold it superfluous for now to present many more of the cases from which the induction of the laws took place, and on which the development of the general concepts are grounded, in order to give more credibility and certainty to the reader. In addition, the form in which the essay appears here and the little time that I had to work out and present the ideas in an agreeable guise make [this design] nearly impossible.

[9] The first of the questions mentioned, namely, which forces are most universal, or which can be observed in the greatest number of individuals of this organic world, is answered with only a little difficulty.

We[9] separate the effects we perceive in particular organisations into classes according to their similarities and differences, and designate these effects or their causes, so long as they aren't better known, with the makeshift word "forces" [*Kräfte*], and with the names of different forces. As long as the differences between classes are not cancelled out by a higher understanding [*Witz*] and converted into similarities, then the following distinguishable and more general classes of effects, or different forces, can be established for now:

1) Sensibility, or the capacity to receive representations as soon as impressions are made on the nerves or elsewhere;
2) Irritability, or the capacity of some organs – above all the muscles – to contract on the occasion of excitement [*Reize*] and produce movements;
3) Reproductive force, or the capacity of organisations to produce and assimilate beings similar to the original in part or in whole;
4) The force of secretion, or the capacity to repeatedly secrete dissimilar materials of a particular kind from the liquids[10] in particular places;
5) [10] Force of propulsion*, or the capacity to move and arrange the fluids in the fixed parts in a determinate order.

If one does not limit one's gaze to effects that are most immediately obvious†, then one has to accept these two latter forces as distinct from the previous [ones], even if the [11] dull understanding sees in them nothing but expressions of irritability,

---

* I employ this word without it being specially selected, because it occurred to me first – another would perhaps be more suitable. In any case, it expresses the phenomenon that is to be denoted here, and that consists of an onward-pushing [*Fortstoßen*] of fluids. Without the cause of this pushing being sufficiently known (until a *reduction* of the phenomenon has taken place), it is assigned the word 'force' for the time being.

† It is clear that I had plants foremost in mind when establishing the concept of propulsive force. For what is also said – primarily in Edinburgh, and newly asserted by Girtanner[A] – about the irritability of their liquid and air vessels, was not so much [said] from the observation of the matter as a presupposition they believed themselves compelled to make. As much as I refer to the phenomena of the movements of liquid in plants, I just as little restrict myself to establishing the concept in relation to them alone. In addition, in those animals provided with hearts, the irritability of the heart and the – entirely unproven and not to be assumed – irritability of arteries do not suffice as causes of the activity of the liquids. Pathological phenomena, the absorption and departure of fluids in the lymph vasculature in which the recently-mentioned muscle fibres believed necessary for irritability cannot be observed, speak here for these phenomena as of their own class: unexplained phenomena that are at present not reducible to irritability.

as Haller[11] found in his followers and pupils – an inert deadweight, so burdensome even to the great man himself.

It is through the universal forces enumerated here – and the various branches of the faculty of representation that can be distinguished, but whose investigation will here be put aside – that, for the most part, nature obtained the great result of the life of the organic world; the relation of these forces are now to be considered.

If the topic is the relations between forces, one must first decide on the scale according to which they can all be measured and quantitatively compared. As long as we have to go without a universal scale which could specify to us the intensity of these forces, then, all things being equal, the number or frequency [12] of simultaneous effects, their multiplicity, and the amount of the resistance the other forces oppose to it, or permanence of effects, in, for the rest, equal circumstances, all appear eminently suited to filling that position.

The fact that, in these respects, a different proportion in the forces of different organisations occurs indicates a passing comparison of the phenomena we perceive in animals and plants; but what will now be shown in the most appropriate way are the kinds of more detailed results given by the use of such measures for the proportions of such forces, and which laws can be drawn from these measurements for changes in proportion in the series of organisations. That is, [these results and laws will be derived] such that, at first, each particular force will be kept isolated from similar ones in other species, and only thereafter will [it] be observed in combination [*Vereinigung*] with other forces, in the way they are found in the individuals [themselves], compared to their combination in other individuals.

Of the aforementioned forces, sensibility earns first place, because nature [13] deals with its distribution among the organisations most sparingly, as it does with the best of what it wants to scatter among the masses.

If we appraise its existence – which we can no longer know immediately in a being [*Wesen*] outside us – according to the existence of organs that resemble our sense organs and nerves, and according to movements that constitute for us the usual accompaniment to sensations, and [if we] determine their multiplicity according to each of the organs and movements, then, with an overview of organisations, we are at once presented with the remark: in the series of these formations, the ability to obtain manifold, distinguishable classes of sensations is gradually more restricted from humans downwards. The sense organs are, in fact, gradually lost in this series, and the movements finally obtain a regularity that no longer tolerates representation as their accompaniments and initiators.

In quadruped animals, and in birds, snakes and fish, all the sense organs, known to be distinct in us, are still present in great perfection*, despite the manifold simplifications that these sense organs received in the latter animals in particular.

---

* That considerable changes of degree in the combination and perfection of the building of the sense organs is also found in these higher animals is sufficiently shown in the research of Geoffroy, Camper, Vicq d'azir and Scarpa[B], particularly in the hearing organ, so too is a simpler view of the visible outer part of these sense organs stays the same. Since here more

In the class of insects, the [14] hearing organ – usually the hidden entrance for the minds of others into the I – has already largely disappeared, and still more generally has the olfactory organ in this class. And even if the eye appears here to have developed more numerously and more elaborately than in other classes – although normally [this is] the organ most capable of communicating with the world at the greatest distances, of taking it up into itself, and of shutting it off from our I – here it is in fact unmovable, no longer capable of being covered,[12] cloudy, and only lets through certain kinds of light. Finally, it is [15] almost completely* dissolved as a separate organ in worms, together with the preceding [senses], brain, and nerves, and in the end persists as only one sense organ that is very receptive to the impressions of light, as we can surmise from the irregular, lively movements of these animals. With plants, finally, this susceptibility [*Empfänglichkeit*] to impressions, which is announced through movement, is still present only in very dark traces, and these only rarely – traces that, through the regularity of their movements, indicate scarcely more to us than some analogue to sensation [*Empfindung*].

Thus, the sense organs – by which so manifold a world of sensations is enacted in us – are surrendered one after the other, and lost in exactly the order, such that those that most increase the physical comprehension of the world for us disappear last. If it can't exactly be said that sensations, too, are denied to them with [the disappearance of] the sense organs, nevertheless a greater uniformity of sensations can be [16] inferred, because organs, such as the eye and the ear, ultimately appear so similarly formed to one another,† and the feeling-surface [*Gefühls-Fläche*] has ultimately to [take] all their places, as a single, uniformly-built organ.

On the other hand, with a second overview of the organisations, it cannot be ignored that precisely where a sense organ is lost and the manifoldness of sensations is therefore decreased, all that is attained is a freer space for one of the remaining [organs], and where one stands as less developed, the other appears all the more elaborated. Insects and worms, largely deprived of eyes and ears, exhibit tactile apparatuses [*TastungsMaschinen*], compared to which even human hands, the hands of apes, and what otherwise corresponds to them in the

---

masses of correlating comparisons are avoided so as to prevent circuitousness, and since, in the above-mentioned animals the inner part of the sense organs still stand very developed, so too, when it comes to the lower animal classes, the assertion made here of the perfection of these organs in the higher animals is generally permitted.

* That in mussels, sea urchins, and more simply-formed worms, and in the greater number of species of this class, no brain [*Gehirn*] or nerve is demonstrable, or until now has not been demonstrated, was shown by Monro[C] and in other research on these animals.

† The similarity between the listening organ – as Scarpa has shown of crabs – to the eye of so many insects and crabs themselves is at times so great that one would almost like to ask whether these animal(s) do not hear with their eyes, and see with their ears.

higher animals, must take second place. So too the receded and diminished eye of the mole appears to have spread out all the more in its more refined hands and nose, and, in a similar manner, in other animals the dullness of the eyes sharpens the ear and the [17] olfactory organ. Even the lower animals in which no distinct sense organ can be found any longer show the capacity of their sensitive surfaces to receive impressions of light in a manner often lacking in differentiated eyes.

From both remarks it now appears (since far more could still be mentioned about these facts here) that the law follows: *the manifoldness of possible sensations in the series of organisations decreases* as the fluency* [Leichtigkeit] *and refinement of the remaining sensations increases in a more restricted scope.*

Consequently, the lack of diversity in sensations among the lower animals is, in each case, compensated for by the intensity [*Innigkeit*] and refinement of the remaining capacities. But closer consideration of the phenomena has the result that, by itself, this replacement of deprived sensibility is not complete. Plants, many of the lower animals, and even several of the series of perfect [*volkommenen*] animals have to bear not only the [18] lack of diversity [of senses], but also a lack of refinement and intensity.

As just now expressed, the law is not universally valid, despite the fact that it contains much that is true and really corresponds to the phenomena, for the decrease in the diversity of sensations occurs in a greater measure, [a measure] to which its replacement by refinement and fluency is unequal. How this law of change in the capacity for receiving sensations in the series of organisations could be more precisely amended will be articulated at once with more precise reflections on both subsequent forces.

The domain of the second above-mentioned force, excitability, is broader than that of sensitivity, although only limited to a greater multitude in the series of organisations, so far as its effects are really demonstrable as such.

Because excitability, unlike sensibility, is not an inalienable force[13], although its effects converge with those of elasticity†, its presence and absence in organisations [19] can be settled with greater reliability. Moreover, precisely because it has many and more reliable marks of its presence, a richer field for comparing its effects presents itself here. Nature has established conspicuous differences not only in view of the diversity and fluency of its expressions, but also in view

---

* In the following laws, the word *while* [während] should replace the word *how* [wie], as the rest of the text makes clear. I left it as it is in order to depict the law more strikingly through the customary form of expression of laws.

† If one understands the necessary physical concepts, it cannot only be said that expressions of elasticity and of irritability coalesce, but that expressions of irritability are surely nothing other than expressions of elasticity. And surely this should be said, if one does not also want to relate to the galvanic experiments what one customarily perceives as the component phenomena of expressions of irritability.

of the perseverance and frequency of its effects under otherwise equal circumstances.

In warm-blooded quadruped animals and birds, soon after the detachment of the torso from the sensorium, and after the separation of individual limbs from the torso, all traces of excitability, otherwise expressed in such a lively manner in these animals' muscle contractions, are extinguished.

However, in cold-blooded animals the situation is completely different; here all expressions inhere almost indestructibly in the organs. [20] Frogs with heads hewn off hop about as if an unnecessary burden were removed, and tortoises continue to move for many days with their hearts ripped out and heads removed.

It would be possible to make similar remarks about fish and insects, because one has seen the feet of many spiders live on and move when separated from the remaining body for seven days, and in the class of worms too, examples of enduring irritability are no rarity, as little as in those plants where real excitability is generally found, since even the little leaves of the hedysarum attached to the torn-off twigs and stems*, as well as the stamen of the barberry, continue their movements, or allow excitement again to awaken them. And consequently, it appears, therefore, that the duration of irritability and its independence from the other systems of the organism in the series of organisation from men down increases rather than decreases.

If one looks around now for [21] the remaining characteristics through which excitability expresses itself in these organisations, then it becomes apparent that some of those which reveal so striking a persistence of irritability are precisely those nature has endowed with separated muscles, or irritable organs in general, in a disproportionately lower number. This is the case in plants, in which usually only some few organs express irritability, as it is in mussels, in which the number of muscles that are distinguished from one another is so often limited to two or three.† In others, in which this permanence of excitability befits their character, if nature did not permit a deficiency in the number of muscles, it arranged these numerous organs in a far more uniform position and orientation. Thus, the four thousand and sixty-one muscles that Lyonet found in the field caterpillar are far

---

* The observations on which I draw here are described in a splendid essay of an unknown author in *Goth. Magazine* 6 B 3 st: "Description of the movements of the Hedysarum Gyrans and the effects of electricity on the same".

† I refer to still unprinted anatomical descriptions of these animals; but even [that research] which has become quite well-known says the same thing.

less diversely disposed and much more similar to each other than the far lower number found in humans.*

The same remark [22] applies to the numerous muscles of fish. In addition, we shouldn't fail to notice that most of the animals that display such enduring irritability are also usually slower in all or particular movements. The movements of most amphibians are lethargic, and their hearts beat – like those of fish – more slowly than warm-blooded animals.

Finally, it appears that most animals with more enduring irritability – if nature neglected them neither in the number nor in the disposition of their muscles, nor in the speed with which they can express their effects – are precisely those animals whose multiplicity of sensations (as in the above) was already noticeably limited.

We see the opposite of the above [characteristics] in the warm-blooded animals, whose irritability is so transient. With them, nature either amassed the muscles in greater number, or, at least, laid them out more diversely, and thereby afforded them a greater diversity of expressions, or compressed many expressions into the same moments of time.

[23] Ignoring other coexistences† that from another perspective might seem more noteworthy, the coexisting phenomena presented here give rise to the following law:

Estimated according to the permanence of its expressions, irritability increases as the speed, frequency, and diversity of the same expressions, and the diversity of sensations, decreases.

In this manner, deficiency in duration of irritability in the different organisations is balanced by the frequency, multiplicity, and speed of its expressions, and by more diverse sensibility. And the above law, by which sensibility changes, has now obtained part of the amendment it needed.

But within this law that changes in irritability follow, one soon perceives that it also requires a new amendment for itself, because the proportion in which the permanence of irritability increases is often more or less exponential than [24] that in which the diversity of expressions of irritability and sensations decreases.

Thus, the duration of excitability in mussels and other animals, and in plants, is (despite the impoverishment of diversity) lower than with amphibians. In addition, in those same mussels and plants, sensibility is reduced almost to nothing, and, ultimately, most plants lack sensibility along with irritability.

Coexistences of other kinds must also be sought here – ones that give closer and more universal determination to both the law by which irritability changes in the series of organisations and, at the same time, the law by which sensibility and irritability disappear. This can now occur, at least in part, by way of a presentation of a few remarks about the third of the previously-mentioned forces.

---

* To be convinced of this, one need only turn to the table in Lyonet's work, or to the muscle table as it readily presents each sliced caterpillar.

† To which they belong with such differences of irritability of the animals [that] in part parallel differences of the medium in which they live, etc.

The *reproductive force* is by far the most universal, and is lavished in the greatest measure on organisations, and one might even want to name it the proper and characteristic [force of organisations] that differentiates them from the other productions of nature, if it could not, as in the previous cases, be sufficiently demonstrated that it too [25] has sprung from inorganic nature, and can be derived from forces internal [*einheimisch*] [to the inorganic realm].*

Because it is universally found in organisations, a greater diversity in the types of its expressions is therefore to be expected. Actually, one must be astonished by the richness of shapes [*Gestalten*] that this force gained at the hands of nature in its different formations [*Gebilden*]. Here it appears in a tremendous bodily mass, there in little dots that our eyes can barely touch or delimit by light. Here it appears in one form eternally, there in the form of a changeable fairy. Here it persists for centuries unworn, there its effects are contained in the passing of a moment [*Zeitfluxion*], here it declares its wealth in millions of new individuals, there it is one that laments its poverty alone. [26] Here it always turns up again with the recurring sun after a brief absence, there, after a single farewell, even the sun can no longer move it to return. Here it resists every destruction and appears always again the same, there a breeze is able to blow it away. Here it is restricted to one place, there it is omnipresent and everywhere equally powerful in other bodies. Here it is as fast as light, and there it creeps lazily along like a salamander and years are needed just to convince us of its existence.

If one doesn't let oneself be bewildered by the multiform [nature] of these phenomena, then this tremendous diversity can be fitted into a few very simple laws, which a brief overview of the decisive facts will immediately develop.

The number of young that warm-blooded quadruped animals deliver after pregnancy stays between the extremes of one and fifteen†, the latter already very rare. Both extremes are higher in the class of birds, and particularly the greatest extreme of fifteen is not only immensely increased but is also more frequent.

[27] With amphibians, species are to be found in which the greatest extreme reaches up to one hundred thousand, whilst their lowest is close to the highest of the mammals, and, in the class of fish, it seems it became far easier for nature to produce many copies of creatures which express little soul; by counting and weighing the roe of these animals, one has found that there are millions in some species. And even in those where nature gets to work with less lavishness on average display a greater fertility than most of the previously-mentioned animals.

But in the lower classes of insects, worms and plants, however much the number of new individuals produced at once surpasses the number in mammals, birds, many amphibians and fish, nevertheless this number is for the most part lower

---

* In so doing, one ought not to bring to mind words such as *attraction, affinity*. How far the validity of this extends will be better shown elsewhere.

† One certainly has examples in particular individuals of species that generate around 19 or 20 young at once. The rarity with which this happens, and only in particular individuals, in the repetition of the business of reproduction allows these numbers to be wholly disregarded here.

than the greatest extreme given for the amphibians and fish. Particularly in plants and worms, the lowest extreme is tremendously small, and the difference between the extremes is greater than those of every other class.

Thus, it now appears that the number of new productions formed in a specific place of the body at the same time increases in a series from [28] men downwards, after some noticeable fluctuation between diminution and proliferation, both in particular classes and in comparatively greater accumulations.

If we now attend to the places in this series where it every time contracts and expands again, the following remarks arise: the four-footed, higher animals, in which so few individuals are produced at once, are also just those in which the mass of their body, in which they appear after birth and that they consequently grow into, is larger than in the animals of other classes. Insects and worms again stand behind even the birds in this respect, and this applies equally to the amphibians and fish in relation to birds.

Further, of the four-footed animals the smallest are again the most productive, the largest, by contrast, are least fecund. The rat produces ten to fifteen young, whereas the whale produces [only] one. From these coexisting phenomena follows the law: *the reproductive force – evaluated according to the number of new individuals produced in a specific location – increases as the size of the producing individuals, or more universally, the individuals produced as they appear after birth decreases.*

[29] In addition, there are other phenomena at hand that should not be neglected, because, from these other phenomena, the law obtains closer determination and more universal validity.

Precisely those animals that show a lesser productivity are also those that have a more complex bodily structure and those in which the young appear more developed after birth. Further, they are the ones that nature needs most time to produce. Nature occupies itself for two years with unproductive elephants in order to form another, while it needs only a few weeks to form a rat. Finally, those animals in which less use of productive force takes place at once are also those that can continue to expend it over longer times.

The butterfly, most insects, and flowers decay as soon as they have transferred their existence to another.* By contrast, most higher [30] animals demonstrate greater endurance in their force of reproduction not only by achieving a higher age, but also by frequent repetition of the business of procreation.

Given these remarks, a more precise and more universally valid determination that the previous law was able to achieve can be formulated as follows:

---

* Aristotle already noted exactly this phenomenon in cuttlefish too, at least in some classes of this bizarre animal species, as Prof. Schneider[D] has shown in his commentaries on the passages in Aristotle pertaining to this species, with an apparatus of so many natural, historical and philological findings, so seldom united elsewhere, in *Sammlung vermischter Abhandlungen zur Aufklärung der Zoologie und der Handlungsgeschichte.*

The more the reproductive force expresses itself in the number of new individuals in a certain place, the lower the mass of the body of the new individuals, the more simply the body is built in which they appear, and the shorter the time used for its formation in the body of the parent, and the less sustained the force itself – or, at least, some of these attributes are found in a lower degree.

Although this law encompasses so many facts, the exceptions are not few, for fish and [31] many amphibians reach high age despite expressing their fertility at once, indicated each year by a tremendous amount of new products. If one wanted to invert the law, it would remain unexplained how such small insects and worms, taken as a whole, would have less fertility than most fishes and many amphibians.

A closer consideration of the animals of these latter-mentioned classes reveals many of the exceptions to be merely apparent, or, at least, as leading to a newer, although more restricted [*spezielleres*] law.

It is striking that precisely those insects that have less productivity on average than fish are also those that demonstrate the most striking metamorphoses or repeated generations of their own body from themselves. Further, it is more striking still, that it is those worms in which (in relation to fish) some restriction of productivity is observed that express artificial reproduction[14] and natural partial reproduction to the highest degree.

Trembly inverted polyps[15] and quartered them and they replenished themselves again[16], Spallanzani* took [32] heads from snails[17] and better ones appeared in their place; and Diquemare† discovered the same wonder of polyps in sea anemones[18]. Further still, it is striking that precisely the less productive of the amphibians – namely snakes and lizards[19] – are also those in which, like in fish, the growth of the body knows no bounds. On the other hand, some of the amphibians in which the growth of the body is more restricted distinguish themselves either, like frogs, by metamorphoses, or, like most newts[20], with unusual expressions of artificial reproduction in addition to metamorphoses. Bonnet[21] removed salamander eyes and feet and, after a little time, they replenished themselves afresh, albeit somewhat imperfectly.

It is still noteworthy in the end that, in exactly those mammals and birds where the least productivity prevails, the newly produced individuals are most differentiated in their genitalia. By contrast, [33] in all other classes – that display metamorphosis, unrestricted growth, or great reproduction of parts as is manifest in many insects and worms, and so on – both sexes become more similar or disappear altogether.

Taken together, these remarks show that many of the phenomena that do not conform to the preceding law can be united under the following new one: the less the reproductive force is expressed in many new individuals, the more it is

---

* MS. *Risultati di esperienze sopra la riproduzione della Testa nelle Lumache terrestri* in the *Memorie di matematica e fisica della societa italiana*, tome I. p.581

† MS in *Philosophical Transactions* vol. 63 & 65. *An essay, towards elucidating the history of the sea-anemonies*, by A. Diquemare.

expressed either through metamorphoses that the body withstands, or through unusual artificial reproduction, or both together, or by indeterminate growth, or by greater deviation in the newly-begotten formations.

Certainly, many exceptions also remain here. Some of which – such as the fact that in many amphibians, tremendous productivity and metamorphoses, indeed even uncommon artificial reproduction, subsist alongside each other, and that in fish unlimited growth and unusual procreation often coincide – have, like similar exceptions that occurred in the aforementioned law of irritability, to be explained by other laws from the nature of the medium in which it lives, its [the medium's] temperature* and other conditions. [34] Others, by contrast, lead to a new law in connection with precisely those exceptions just mentioned above.

Precisely in the lower classes, where, in every respect, irritability and sensibility are only meagre remnants, in plants and many worms, very often all the aforementioned varied expressions of the reproductive force show themselves united in a very high measure. Plants replenish what they have lost anew, issue an enormous number of seeds, and grow into such a mass, which [that of] the largest animals must lag behind. And, in mussels, one finds examples of a similar union of different expressions of this force to a very high degree.

By contrast, in the higher animals, where sensibility is still present in greater diversity, the expressions of this force are restricted to fewer, [35] or at least united in no such high measure, and therefore the universal law results:

The more united all the kinds of expressions of reproduction are in an organism, the more the faculty of sensibility is excluded, and the more even irritability gives way.

A closer consideration of the remaining forces mentioned above, namely the forces of propulsion and secretion, which has here been put aside because of its enormity, could – because these forces permit remarks similar to the earlier ones – confer still more precise determination to these newly-discovered particular laws.

Thus, the laws by which particular forces in the series of organisations are altered have now been supplied. The question of what can be drawn from these single laws [to constitute] nature's scheme for changes in the relations of these forces taken together now permits this brief response: in the series of organisations, sensitivity is gradually displaced by excitability and the reproductive force, and, in the end, irritability too gives way to the latter: the more the one is increased, the less there is the other, and sensibility and reproductive force tolerate one another least. Further, the more one of these [36] forces is cultivated on one side, the more it is neglected on the other.

On the other hand, it is no less discernible that despite these compensations, not only do the individual forces decrease at the expense of the others, but also the

---

* How much the warmth of the medium the organisations are accustomed to determines the types of expression of the reproductive force is shown not only if one compares the fertility of the individuals of a species that are placed in different climates, but also, if the different species of one lineage are compared – thus very similar animals – in which one of these difference in their habitat occurs.

sum of forces decreases in unknown proportions – and this can be explained neither by the medium in which the animals live nor by other circumstances.

Consequently, this relation in the series of animals is changed according to very simple laws; the simplicity of these laws, which flow into such a tremendous diversity, must stand out even more, if one considers that precisely these laws, by which the forces are distributed to different organisations, are also precisely those by which the distribution of forces to the different individuals of each species occurred, and, indeed, to one and the same individual in its various periods of development. Men and birds are also plant-like in their first state; the reproductive force is active in them.

Later, in the moist elements in which they are then living, their irritability increases; moreover, the heart of these [37] animals is indestructibly excitable;* and [it is] only later on that one sense [organ] after the other develops in nearly exactly the ascending order in which they emerge in the series of organisations, and what was previously irritability ultimately develops into the capacity for representation, or at least its unseen, most immediate material organ.

Nature must appear still more uniform in those laws if one finally considers that the same laws are observed even in the distribution of forces to a particular organ at different times. Now, in the female uterus, the force of secretion is active, a little time later it changes exclusively into reproductive force, and finally this too is replaced by irritability.

Even without getting into everything that can be deduced from these laws, and without developing further what can be deduced about the cause of the appearances, which I have so far distinguished under the names of specific forces, or, indeed, even [without developing further] what can be concluded about the efficient cause of the distribution in the different organisations – nonetheless, we cannot avoid remarking on how rich these laws are in consequences of this kind.

[It is] not only [38] in the series of different organisations, but also in individuals of the same species, that, with the disappearance of one force, another emerges – without a more universal, coexisting phenomenon being noted here, for it happens within single organs [too]. The vanishing of one can therefore be regarded as the cause of the emergence of the other, and so one can infer a common† cause, if the cause of these appearances is the issue, and were the material cause known in one case, the cause in question could boldly be deduced from the other.

---

* This assertion is grounded on observations I have made of the hearts of little chickens, ducks, and geese, removed at the point at which they were half-developed.

† One obtains the same result if one handles other species of premises – in fact, the species one may use are many. I have presented one of them in a couple of letters in the newest issue of Gren's journal of physics[E] together with the conclusions drawn from it. The conclusion alone, as it is made here, thus stands supported on various sides all the more surely.

Indeed, because the distribution of the forces in the series of organisations follows the same order as the distribution in the different developmental states of the same individual, so it can be deduced that the force by which generation occurs in the latter, namely the reproductive force, is consistent in its [39] laws with the force by which the series of different organisations of the earth are called into existence. And because it is precisely the lower classes in which individuals are so numerous – that are produced the most numerously in species – it is therefore still more permissible to accept that the force by which the series of species are generated according to their nature and laws and [the force] through which the different states of development are brought about are one and the same. This could also actually indicate, if this were the place to further develop the idea, that one could be led by carefully selected analogies to [the point of] assuming such a material cause as an explanation of the developmental appearances that could be imagined as at work in the first production of organisations on our Earth.*

Setting aside these and other consequences, I will now take the more obvious consequence in hand, and answer without further ado the third of the [40] above-mentioned questions: how can the course and continued existence of animated nature be explained from the laws that specify changes in the relations of forces of organisations?

Because when comparing the distribution of forces in all organisations, no very marked dissimilarity was produced, and the more excellent force of one was almost always offset by such an excellent force of the other, which, in the case that the first worked towards destruction, equally necessarily entailed preservation, then the continued existence of the organic world must have emerged from this equilibrium of the destructive forces and the forces of preservation, and so too its course, in so far as the mutually balancing forces are always different and express themselves in different ways.

So it happened, therefore, that plants, robbed of sensibility and irritability, repel all destructive forces of the animal kingdom with their reproductive force, and so it happened that, despite the depredations that all classes of animals venture against the worms, and despite the persecutions that they themselves perpetrate against each other, this class – by its indestructible irritability and reproductive capacity – withstands each attempt to annihilate it. So it happened that precisely the higher animals, whose mass pushes them to greater destructiveness, [41] are by no means supported in this destructiveness by a great number of individuals; and so it happened, ultimately, that animals whose more swiftly and frequently expressed irritability and greater sensibility urged them towards predation, were, because of the small number of their individuals, less able to do so. Briefly then,

---

* I must again refer to the first remark of the essay and my investigation of developmental phenomena implied there. In addition, there isn't space to back up what was said here with further evidence, even though that may lead people to suspect me of being overzealous.

destruction by one force was thus regularly repelled by another force, or limited by a different side of the same one, and thereby the species was at least preserved and the individual only carelessly abandoned if its individuality does not count for much, as is the case in plants and worms, in which greater dangers and greater incapacity to feel pain are equally present.

As much as the continued existence of the organic world is in this way founded on those laws, nevertheless it cannot be denied that it is exactly those laws that, on the other hand, also permit the question as to whether here and there an over-preponderance in the destructive forces does not actually arise, and whether species whose forces of preservation are inferior must not eventually themselves pass away?

Moreover, since, in addition to the greatest possible balancing [*Ausgleichung*] of all forces with one other, a gradual decrease of the sum of forces in the series of organisations was also noted in the plan of nature, so the emergence of [42] one such over-preponderance is thereby more than merely possible; and, if we now only stick closely to what observation actually tells us, the places in this organic world where over-preponderance emerges – and where it works most forcefully – in fact appear to become somewhat more precisely determined.

This is the condition of many species of mussels, who, by scant compensation in their forces, and by the wealth of forces that nature gave higher animals for [the mussels'] destruction, are placed on such a dangerous peak that one can assume with likelihood that the decline of one or other species is gradually occurring, and has perhaps already often happened. What so many higher animals are capable of when it comes to some mussel species, humanity and its species are capable of doing with respect to so many other, and even higher animals.

With the rationality that turned up in humanity's organisation, humankind obtained the capacity to freely alter (within certain limits) the relation of the other forces that it has in common with the other animals. It created microscopes and telescopes for eye and ear, and thereby heightened its sensorial capacity, and who knows whether further similar improvements may not be applied to smell and touch.

To increase its capability of movement, [43] it coerced other animals, fire, and wind, to lend theirs to it, and, through these changes that it undertakes in the relations of its forces, and through its greater capability to endure each of these changes, it obtained, with its species, a decisive preponderance over most other animal species and their forces of preservation.

Humanity also actually displaced other species of animals, transferred them to small areas as strangers while it took the rest of the Earth into its possession; indeed, it enslaved entire species, and it is more than merely probably that it will still compel many more to completely exit the stage in order to make space for itself, as one organ replaces another in the great machine.

In this way, the course of development of the organic realm is founded along with the way that different forces themselves are distributed. And for this reason – to take a quick backward [44] glance – in the same way that light appears split into different rays and these rays are mixed into infinitely different proportions, the smallest organ, right up to the most complex and immense machine, is set in motion by a single force, a force which perhaps was originally awoken by a light whose daily support it still enjoys.

* * *

I conclude this treatment by permitting myself only a brief return to where I began and which I address, above all, to you my friends. In the periods of its development, the human spirit also displays changed proportions of the forces united in it. Sensations are the playmates of children, fantasy the choice of adolescents, and understanding the true companion of humankind. One arises from the other like seeds from flowers, and flowers from branches, and with the developed seeds of one the blooms of others wither. However, such a change in the forces of the human spirit is not only noticeable in the different stages of life. In the same person, changes of circumstance present these [changes of forces] with their development compressed into short periods.

Today a narrowed sensory world gives the realm of his imagination [*Fantasie*] greater extension and fixes the edifice of his understanding, and tomorrow this imagination returns to its prior boundaries [45] and yields to the senses their domain once more, and the happiness he feels in one cancels the happiness [felt] in the other.

Indeed, nature has presented precisely this development of the human spiritual forces [*Geisteskräfte*] from each other – which can be seen following one another in the same human – in different humans at the same time in isolation.

The northerners compensate for the poverty of their strip of land – which presents so little variety for the entertainment of their senses – with calm imagination and cold understanding. By contrast, the occupants of the hot zone, who nature allows to look upon a richer sensory world, must do without this compensation. Here too a similarity between the laws by which the organic forces and the forces of the human spirit change their proportions is notable. Now, just as through the former, duration is ensured in the realms of organisations, in the same way, through the latter, now too an enduring enjoyment [*Genuss*] of fortune is assured. With this capacity to replace each loss in another respect, [humankind] could spread its species over the earth, and it is with this capacity that each individual man, even though so numerous, still now navigates the wide sea of life safely.

The kind of enjoyment[22] can change, but not the inner enjoyment itself. With every privation that the human spirit experiences on one side, it [the human spirit] supplements itself, like the polyp, on the other. A flower cut off [46] only brings forth a new one.

But as powerfully equipped with the capacity to compensate as the human spirit is in its struggle with all the rest of nature that works against it, it is just as certain that this capacity is only an inclination that first requires care and nurturing if it is to display higher efficacy. For, like our individual forces, it too can be repressed. The enduring influence of adverse circumstances could dissolve the most beautiful rays of our spirit, and thereby give to lower forces a lasting preponderance [*Übergewicht*], and, in this way, we would lose the character of our species – that capacity of compensation, or in other words, the capacity to flourish under all external circumstances.

Thus, eternal thanks are due to those who – since they do not have the power to steer fate – cultivate that tendency in us through which we confront it [fate], thereby striving to make us wise and happy. Eternal thanks therefore to the illustrious founder of this academy, who, within this group, brought together so many means for achieving that goal. Eternal thanks to the regent of our homeland, for he too strives with such tireless determination to reach the same goal in this greater circle as we do here in our smaller one, and to protect the fortune of the individual against adverse conditions. And may he receive these thanks today as the purest offering that we can bring him: most spirited wishes for his life and everything that can bring him joy, from all of us.

* * *

[*What follows is an extract from an early, unpublished version of the 1793 Speech*[23]]

[63] For each of the just-mentioned organs from which I [NATURE] assembled the individuals bearing those bodily forms, I have established that – brought into connection with the rest – it experiences changes in each infinitesimal part of time; I have adapted the changes in one to the changes in all of the others and united them in a system of simultaneous and consecutive changes in such a way that each becomes, according to your manner of speaking, reciprocally cause and effect of the other, and is so interwoven that all that could remain as a task for your Lesage[24] would be just to say where I began and ceased to weave.

And yet this is only a trivial part of what I do – an [64] individual, momentarily living organism. Further still, I lead each of these living individuals onwards again for a greater or lesser stretch of time and, at each point of this course of time. the system of effects, which you call its life, and the system of organs, which constitute its organism, change – one emerging from the other as from a cause; childhood, youth, old age and death lead on one from the other, and, if I could only lead you out of time and space and follow that path with you from where it departed from your system, then you would probably have to admit that to die – which is a change that you customarily do not highly esteem – is no less artificial than to live.

In each of these conditions the effects of individuals stand connected to the effects of other individuals of the same species in a greater system of effects: the infancy of one depends on the maturity of the other, and the youth of one on the youth of the other. And this connection is so tight that you should believe, according to your manner of speaking and representing, that I had interwoven the nerves of an individual with those of the others into a web, and the impressions of one would be felt in the sensorium of the other.

And the result of all this is that individuals too live on with the species [*Gattung*], and that, despite the dispersion of individuals, the species lives on. I let this [65] new great system of effects – the living species – also progress along a path of development in greater periods of time, as you can most clearly perceive in your human species, and, despite the fact that the history of your race [*Geschlechts*] has permitted you to see only a small element of this path, so now the joyful youth dawns from the infancy of your species. Whether one day I will let your species too (like individuals) be replaced by another, newer species [*Gattung*] – you need no information on that matter for now.

Finally, the effects of the individuals of a species interconnect with the opposed effects of individuals of all other species in a system in which one is cause of the other – and the result of this new system of effects is the life, not of individuals in common with the life only of the individual, but [that] produced out of the life of all others, the life and contents of the great machine which, I said before, is made up of at least roughly 7 million cogs, of which none is immediately connected to the other in space, but are only chained together through time. This entire machine is adapted to inorganic nature and its changes in such a way that it is as if they created each other. Now, if some of you take me to be so weak and, out of pious enthusiasm, were to burden me with the claim that [66] not even one of all these cogs could be taken out to be repaired without the whole system collapsing, then know this, and this I say to you myself: I have already on occasion annihilated stars above, and you only learn of it after a few hundred years. I have exterminated animal species from the earth – but what happened: a different species arose, and, in brief, I lead this great machine of the organic realm along a path of development which, like the developmental path of the universe, perhaps like the developmental path of the individual, could be represented to you with the image of a parabola that never closes in on itself, because I do not clearly show you even one element of this path.

And if now some among you, whose courage and daring I must at least admire, said directly to my face that I had no intent with all these artificial placements of phenomena one after another and next to each other in time, and each effect and consequence were not an end [*Absicht*] I wanted to achieve – I want to say it myself: I had no intent – you will nevertheless have to admit, after all, that each chain of effect and cause looks like a chain of means and end to you (I want to appeal to your vanity here and treat you as an equal), and it is very conducive to your reason to accept such a chain, even if you just wanted to survey things better and more easily [67].

Nature could go on speaking in this way much longer, and it had still not even touched upon what is most important, I mean the wonders of the power of representation. And if we dispose of the slight boastfulness that was discernible here and there, then we would have to at least admit that it could convince us of the truth of what I said at the beginning.

Translated by Lydia Azadpour

Translator's Notes

1   This speech was given on 11 February 1793 in honour of the birthday of Carl Eugen, Duke of Wurttemberg, founder of the Stuttgart *Hohe Karlsschule* where Kielmeyer had

taught since 1785. It was the only work of any substantial length that Kielmeyer could be persuaded to publish during his lifetime, appearing later that year. For more contextual details about this speech, see Chapter Two and the interpretative chapters which follow. This translation follows the facsimile of the 1793 Stuttgart publication of *Über die Verhältnisse der organischen Kräfte unter einander in der Reihe der verchiedenen Organisationen, die Geseze und Folgen dieser Verhältnisse,* reprinted with an introduction by Kai Torsten Kanz (Marburg: Basilisken-Presse, 1993). The numbers in square brackets refer to the page numbers in this original edition. All footnotes are Kielmeyer's own.
2. Carl Eugen, Duke of Wurttemberg (1728–93) founded the *Hohe Karlsschule* military academy in Stuttgart, which became a university in 1781 but was closed in 1794 after the duke's death.
3. 'Phenomenon' is used to translate 'Erscheinung' throughout the texts that follow, although, of course, 'appearance' is equally possible. Both possible translations hopefully make clear Kielmeyer's closeness to Idealist discourse at this juncture.
4. Kielmeyer's use of '*belebte Natur*' refers to all living organisms (bacteria, fungi, plants, animals, etc.).
5. By '*die Organisationen*', Kielmeyer refers not only to objects bearing certain structures, but also the result of a process and the activity by which an organism is formed, see, for example, the Grimm-Wörterbuch entry on this term.
6. Pierre Lyonnet (1706–89): Dutch natural scientist and illustrator, best known for his illustrations for Friedrich Lesser's *Theology of Insects* (1742) and Abraham Trembley's *Mémoires pour servir à l'histoire d'un genre de polypes d'eau douce* (1744). In 1750, he released a detailed examination of the muscles of a caterpillar (*Anatomical Treatise of the Caterpillar which Corrodes the Wood of Willow*) to which Kielmeyer refers later in the speech.
7. An allusion to §64 Kant's *Critique of Judgement*.
8. The term '*Verhältnis*' at stake in the very title of the speech is variously translated as 'relation' and 'proportion' in what follows, in order to capture the fluidity by which Kielmeyer moves between indeterminate and determinate connections between forces and organisms. However, whenever 'relation' is used, the reader should bear in mind that it is always, for Kielmeyer, a determinable proportion of forces.
9. Reading 'Wie' for 'wir', following Kanz' facsimile.
10. The term used here is '*Saftmasse*' that is a 'colourful and colourless blood in the vasculature and tissues' (see G. L. Ofterdinger's essay in the 1829 volume of J. F. Meckel's, *Archiv Für Anatomie und Physiologie* (Leipzig, 1826–32), 248).
11. Albrecht von Haller (1708–77): responsible for developing the concepts of irritability and sensibility in physiology.
12. That is, no longer capable of blinking
13. By 'inalienable', Kielmeyer seems to be noting that excitability is available to an external viewpoint, whereas sensibility is something that the individual organism possesses internally.
14. By 'artificial reproduction', Kielmeyer refers to the regeneration of *parts* of the body.
15. Now known as freshwater polyps, in the genus *hydra*. Trembley's experiments were published in 1744 as *Mémoires pour servir à l'histoire d'un genre de polypes d'eau douce*, and this work had been translated into German in 1791.
16. Abraham Trembley (1710–84) – see previous note.
17. Lazzaro Spallanzani (1729–99): an Italian pioneer of microscopic observation, whose research included experiments that contested the theory of spontaneous generation.

18 L'abbé Jacques-François Dicquemare (1733–89): he wrote on topics in astronomy and the life sciences and published *An Essay towards Elucidating the History of the Sea-Anemones* in 1774–5.
19 Kielmeyer here speaks of 'die Landeidechsen' or the common wall lizard.
20 Kielmeyer here speaks of 'die Wassereidechsen' or the smooth newt.
21 Charles Bonnet (1720–93): Genevan natural scientist, insectologist and advocate of preformationism.
22 Kielmeyer here uses the term '*der Genuss*' to signify enjoyment in the sense of 'advantage' or 'use' (rather than pleasure).
23 What follows is an extract from a manuscript bearing an earlier version of Kielmeyer's speech, which is printed in *GS*, 63–7. This translation follows the *GS* edition. In this passage, 'Nature' ('I') speaks directly to the human reader.
24 Georges-Louis Lesage (1724–1803): Genevan physicist whose work on mechanics and contributions to the *Encyclopédie* rendered him the most famous scientist of the day across Western Europe.

Translator's Notes to Original Footnotes

A Christopher Girtanner (1760–1800): Swiss chemist based in Göttingen and Edinburgh.
B Kielmeyer here names some of the founding fathers of eighteenth-century comparative anatomy: Étienne Geoffroy Saint-Hilaire (1772–1844), Petrus Camper (1722–89), Félix Vicq d'Azyr (1748–94) and Antonio Scarpa (1752–1832).
C Alexander Monro (1733–1817): Scottish anatomist, President of the Edinburgh Royal College of Physicians and friend to Petrus Camper.
D J. G. T. Schneider (1752–1822): editor of many Greek texts on natural history, some of which were collected in the cited *Sammlung vermischter Abhandlungen* in 1784.
E F. A. K. Gren (1760–98): Professor of Chemistry and Physics at the University of Halle from 1788 to his death.

Part II

SELECTED UNPUBLISHED TEXTS BY KIELMEYER

Chapter 4

ON NATURAL HISTORY[1]

C. F. Kielmeyer

Bebenhausen, April 21–3, 1790

I. Concept of the object of this science [of nature].
II. Relation of this object to humanity's power of representation[2]
   a) consciousness of this relation, which is science itself.
   b) division of this science according to the variety of objects and the variety of human forces engaged therein.
III. Value, misuse, and use.
IV. Requirements derived from the concept of science for the correct presentation of objects.
V. General enumeration of the objects produced by observation, the necessity of this enumeration – the wheels of a machine – to become acquainted with the individual standing under the influence of others.
VI. History of the Terrestrial Sphere.
VII. Universal properties or those applying to it as a whole.
VIII. Survey of its major virtues.
IX. Phenomena in the fleeting stratum.
X. Phenomena in the fluid stratum.
[XI.] [Phenomena in the solid stratum.[3]]
[XII.] [Commonality and cooperation of the strata.]
[XIII.] [Diverse processes in the operations of natural history.]
[XIV.] [Attempt at a graduation of the productions of our terrestrial body.]

[212] [*I.*] Before we allow ourselves, following custom, to praise the science on which we are to lecture, it would be more in keeping with the rules of a lecture to establish a concept of the object with which science deals. Only then, and by a more exact survey of that very science, can praise be fruitful for you and, owing to hopes of success, I for my part may undertake it with greater warmth and interest.

Words that initially applied to designate sensible objects and their changes will later be transferred, by virtue of higher human wit, to the abstract concepts bound up with these objects (the human thirst for ever new matters of conversation is bound up with the drive to express all the capacities present in it, [and] is soon teased out from it), and thereby obtain a wider meaning, losing their original which must, if an all-encompassing explanation is required, be subordinated to the broader. The disproportion of the number of words relative to the infinite mass of possible and ever newly-emerging concepts also enables the meaning of the words to be extended to abstract concepts that stand in some relation to the foregoing, but which are indeed quite dissimilar to the original object. In this way, and all the more the [213] more often a word undergoes this fate, the meaning becomes so broad that it becomes almost impossible to collate what is similar in the subordinate concepts into a single commonality.

In this way alone, it can eventually happen that words that initially hold a definite concept in themselves, lose their concepts through widespread use by those who should first assess the similarity of concepts but are perhaps too lazy, and leave them as shells, like the mussels and the fossils of sea creatures, whose originals have perhaps been entirely torn from us by some kind of violent revolution.

This is how things are in modern times with the word *enlightenment* [*Aufklärung*]. Its various adherents are content to designate contradictory things with the same word – and the word *nature* [*Natur*] has indeed had a similar fate, considering the various concepts to which the word is deemed appropriate.

The concept founded on the derivation of the word *nature* is probably the first, in that it contains emergence, becoming, birth [and], within it, giving equal opportunity for applying the word to many related concepts. In consequence of the laws by which humans coordinate and call forth concepts, the thought of the *actus*, the emergence of a thing, may now induce the *actus* that is the cause of this alteration, now [induce] its kind containing the simple historical presentation of the phenomenon, [or] now [induce] the laws according to which this alteration proceeds. [214] And all these things are indeed actually designated by the word 'nature': [it designates] both the productive cause of things (*philosophy*) – by virtue of the association of effect and cause – and the things themselves (namely, the product), insofar as they become bound together in time and space, or, antecendent [to them], things for themselves as the effect or interrelation of the laws to which the fate and connectivity of things are subject, insofar as they are grounded on the determination of the highest cause (*physics*) or the developments and fates that arrest the course of a thing (*physiology*).

Insofar as humanity, beginning with the developments and fates it discerns, can posit an essence or abstract the essence of a thing from such developments; or also, insofar as the entire future is grounded or is, at least, related to the beginning of a thing (which is often not the actual, first beginning, however), it is also acceptable to apply this name to the essence, the properties contained in a thing. And in this respect alone, though not according to the original concept of the word, it is also permitted that it be used to designate the sum of god's properties, which are not grounded in a beginning – a difficulty in which man is caught when he applies human speech to things that are outside his sensible world.

The naturalist or better, the pantheist, who finds in the product itself the cause of its emergence, [215] covers cause and effect at the same time with this word.

[*II a.*] Now, after the highest concepts that the word nature is apt to cover have been enumerated, it becomes necessary to emphasize those which the natural researcher [*Naturforscher*] attaches to it:

Nature is everything emergent or actually appearing to our senses and which is perceived with our outer and inner sense, connected in time and space and apt to follow certain laws.

Since a machine in the broadest sense is a combination of one or more forces (towards a specific end) by the cooperation of which certain results will be obtained, it is permitted, if the larger results are not obvious to us and to the extent that we note interconnectedness, to represent nature as a great machine whose cogs we grasp according to all the phenomena that arise for us.

We usually call the interconnectedness of things, when contiguous with one another, the world. The concept of nature is broader, it conceives, at the same time, things following upon one another – or, as St[orr]* said, the mechanism [*das Werk*], which – when considered as composite – is called the World, [216] is Nature seen in motion.

St[orr]'s definition, that, for the physiogr[apher], Nature is the inmost essence of all appearances contained in the course of the World, is consonant with what I have already fundamentally established.

From the concept that the word nature comprises in things conceived in the course of their development and fate, various adjunct terms have been derived: hence *natural* is what follows from the state of things left to itself, as opposed to *artificial*, which designates an interruption of this state worked by human hands – although surely the distinction here made is entirely arbitrary, since man himself, and his behaviours, belong to nature. Instead of this, an antithesis to *unnatural*, which encompasses every deviation from the usual course of things, would be better, although the statement would be unusual, since monsters also follow from the state of things of themselves and are natural – therefore the medical use of unnatural things is incorrect.

In [the term] *counter to nature*, which signifies a contradiction with the universal laws and properties of things, the contradiction is not merely apparent, as is very often the case with absurd beliefs, but actual, which humankind in *physics* is seldom able to demonstrate, because contrary [217] to nature is at the same time impossible, and in consequence the expression is ill suited to designate the state of the human being in opposition to that of health, and, as one does not say unhealthy in place of incurable, unhealthy would [here] be the better term. *Supernatural* is unnatural, and its ground is ascribed to a force sublime above the world's forces – and is often, for humans, counter to nature at the same time.

---

* [G. C. C. Storr: *Sketch of a result of conversations on the introduction of natural history* I, 1777.]

Since the goal of natural history is to recognize and to distinguish what is produced on our Earth, hitherto generally referred to as natural bodies (amongst which, according to the above statement, are included all things, as they arise by their own accord, sensibly perceptible through our senses), according to their connection to our senses, the means of achieving this end must also be given. In general, observation, comparison and designation would be given as the means to achieve this end. Now, since the object of natural history has been set out in more detail, so too the application of the means to this objective must be defined in more detail.

Since the quantity of objects whose connections to our senses we experience is so great that we cannot commit all representations of them to memory to survey them without loss or deficiency in having to recall them, [218] its representations must therefore become more determinate, fewer, and easily recalled together, whereby the representations of the individual objects can be absorbed and reproduced more easily and more permanently, so that knowing is made easier. Thus, if I wish to impress representations of animals, plants on my memory, I make the less determinate representations of the region and the time in which they grow dependent on these, whereby each time they may be easily recalled in consequence of the laws of association.

Now since, further, there is great similarity amongst objects, and I distinguish them without confusion to impress them on my memory, these less determinate representations which I attach to representations of individuals, are not derived from unstable circumstances lying outside the objects themselves, but by the comparison of representations that are themselves taken from objects. –

If, then, I attach such lesser signs to the representations of individual objects, I thereby sort representations and objects into groups and divide them. Now a sorting of objects according to these signs means an *Order,* a *System,* a *Method.*

In short therefore, the goal of natural history, combined with restricted human abilities, gave birth to method in its [219] broadest sense (not only [this]?, even the kinds of human capacities make it necessary).

[*II b.*] This is evident not only from a general consideration of its origin, but also from the specific history of the latter.

To begin, [method] was descriptions, observations of individual objects, and one merely brought together into series creatures similar in everything but birthplace or habitat. Individuals were gathered together through comparison into kinds and species [*Arten und Gattungen*], and observations of individuals within these species collected. The necessity that all these diverse observations of such kinds remained discrete is evident from the limited number of objects [observed], since in the beginning it was restricted, as it still is in undeveloped nations, to just one, particular area, to nearby habitats, and certainly to objects dictated by need – although no longer. As the amount of observations began to rise, humanity no longer restricted itself to one particular area, to the closest habitats and to the creatures that served to satisfy human needs; humanity's desire to know is no longer determined, as it were, by bodily needs; instead, the feeling of boredom, and consequently active intellect [*Geist*] and thirst for conversation drove the desire to know, making a method all the more necessary. The named kinds [*Arten*] come to be [220] distributed into fields alphabetically and by birthplace.

Since it is grounded, albeit entirely arbitrarily, not on homologies noted in the parts of natural bodies, but only on the similarity of the signs, the first kind of classification is, indeed, only sufficient to evoke previously-held notions and their signs in small number, but is not at all sufficient to discover the as-yet-unknown signs. Moreover, it confuses the mind, which calls forth the things themselves according to the similarities. The second kind of classification, to arrange according to birthplace, is easier to recall, also easier to find later, and is bound up with the appropriate comparison of all obtaining circumstances. It is thus one of the best, in that it is in part the basis of, and partly itself is, a geography[4] – but since the most different things may appear in proximity, and here, as in the skies, a classification must be made, so when knowledge is extended, the classification is no longer applicable, since plants should also be recognised and known beyond their habitats.

[*III.*] The primary purpose of the method was to have a thread that would connect knowledge of individuals and be appropriately aligned for memory. However, if I were able to find this thread while, at the same time, not yet knowing the individual, [221] then not just recognition but cognition too would be made easier.

The threads along which known objects should be arranged in line must themselves be brought forth from the comparison of objects. On comparing several individuals, it was found that in every respect but that of space, which they could not occupy simultaneously, they were congruent with one another or differed only so far as may be hypothesised of things descended from a single original. These individuals or what is common to these individuals, obtained through abstraction, are grasped under the name of species [*Gattung*]. Those alike within a species would be called a race [*Geschlecht*[5]], and all those natural bodies bearing homologies to one another are together held within it. Such abstraction from characteristic traits, and the classification based upon it, will proceed just until the number of divisions is much reduced and easy to survey. Classification according to these levels [*Stufen*], the number of which cannot be determined or will only be determined according to the number of bodies and their varieties, is called system, method. Yet comparison is always engaged with those individual parts alone that show certain relationships in different individuals.

Because this comparison is one-sided, it immediately follows that as many comparisons, and consequently as many systems of this kind may arise as there are comparable aspects of things. [222] The greater the importance of the selected aspects, the greater the dependency of the other parts, their stability and duration, decide the quality of a system, which then satisfies the task of distinguishing bodies (according to these traits). But since a comparison resting on a few individual parts, and due to unequal evidence, as well as to the author, establishes nothing definitive by itself and must, by its law, set together things that are otherwise entirely different, another means would additionally be adopted to remedy this (for as long as no such ground of classification has been found with which everything else, as in e.g. *Didynamia*,[6] would immediately be given). When all parts are compared, the number of homologies should decide on kinship. However, several plants may agree, e.g. in relation to hair, colour, root-orientation etc. and disagree in fewer yet more important [instances], but must subsequently be correlated according to the arbitrary law. Thus, a new difficulty arises for anyone who wants to know and order things as they really are most similar to one another, and does not want to separate from one another things that a glance has already taught him are so similar; he does not wish simply to bolster his memory, nor merely to distinguish by individual traits, but rather to know them in their relation to all his senses and, taking all these relations together, [223] he wants a means to facilitate memory that should at the same time lead him to infer by analogy from a discovery made in one body to others with greater certainty.

[*IV.*] After what has been enumerated thus far regarding the goal of natural history and the means by which we are to achieve it, it is now necessary to illuminate in more detail the application (made actual) of this means to that end, the achievement of the content of the science.

Among the phenomena whose connection to the senses humans experience, those clearly warrant primary attention under the influence of which the others seem to us to stand; those, that is, in which consequents seem to be grounded as effects in causes.

Having observed this order in a lecture on Natural History, it directly resulted from the concept of science announced there as the best, and – if it were possible to observe this order in all that science's achievements (that is, were the relation of time and space with a phenomenon itself always given to us in respect of other phenomena), if we knew their coexistence and their following on from one another – then the historical knowledge of nature would be complete for us and even the [224] theory of these phenomena, physics, would have fewer gaps, although we are otherwise justified in reasoning from 'always precedes' to 'necessarily precedes', to the cause, and, in consequence perhaps, the theory and history of phenomena do not wholly coincide.

[*V.*] Sleep, which in the great part of organised nature follows from the absence of sunlight, [and] the organ of the eye (this very powerful sense that, by means of light, may touch immeasurable distances of stars) that is awakened and endowed, or acquires its effectiveness and use by means of the medium alone, by light, convince us, if nothing else, of the powerful influence of the sun. Moreover, the diversity of the angles by which light's rays strike our Earth determines the greater or lesser amount of heat by which the latter is either elicited from our Earth, chemically precipitated, or emerges anew through bonding with an ingredient of our Earth, or produces the movement that is excited in us by the sensation of heat.

At times such as Spring and Summer, when the rays of light, in relation to the position of the Sun, commonly fall upon us at less oblique an angle, with the greater distance of the Sun itself, the organised nature [225] that, through the preceding season, had sunk into a deathly sleep due simply to the paucity, the acute angles, at which the sun's rays fell, is awakened anew. The saps of plants now move not only through their accustomed channels, but even form new ones. Insects, worms, amphibians are again newly awakened and, as they are cold-blooded, are influenced to a far greater degree than the warm-blooded animals, excepting fish, [for] their medium's temperature guarantees strong protection against air temperatures.

Yet it is not only the awakening of their own nature in this season that depends on this small imbalance, but the production of others of their kind seems also to be tied to this slight thread. Love's drive occurs again in the vast majority of organic creatures at that time; now, whether it is the case that all these phenomena are produced by the heat awakened by the sun's light or by the direct communication of the light itself, the drive to communicate the abundance of this matter to another is aroused. These phenomena, in short, always follow in this way.

If, moreover, we consider that light appears as an ingredient in the majority of the bodies on our Earth that take up light as matter [and] that, for perhaps a greater part of our Earth, the transition from one mineral kind to another, the erosion of our Earth's crust, is to be ascribed to this ingredient, [226] to this alien [*fremden*] arrival, then we cannot restrain ourselves from saying, with *Brockes*:[7]

> All of this together shows me,
> that wonder stems, alone, O sun, all at once from you,
> though this sunshine too
> stems alone from you, O God, alone.

It is additionally clear from this how peoples rich in fantasy, who remain solely with what is representable without concerning themselves with the thinkable, believe they have found the reason for all appearances in the sun, which they are able to view as their Godhead.

The opinion of some ancient philosophers appears less laughable, and, so long as we cannot prove the opposite, never absurd, even after examining this phenomenon. Empedocles, for example, who held that animals and plants are produced by the sunlight on a damp mire.

This entire episode therefore shows us that, if natural history is to be carried out in accordance with this initial axiom, the class of phenomena to which the Sun belongs – the Heavens – encompassing every phenomenon found beyond our habitat, or the bodies similar to our Earth as a whole, must be the first objective of our closer scrutiny, insofar as knowledge of it is historical to the extent that its results[8] are facts.

[227] However, since the path by which one reaches this knowledge is not mere observation, but a comparative observation supported by *mathemata*, and is, therefore, an essentially different kind of knowledge; and since, moreover, this knowledge is the subject of a particular and genuinely historical science that cannot even be separated from the others without silencing knowledge itself, in

respect of its observations as well as its results; then this topic does not belong here because, further, the results themselves teach us by way of relief only of their movements, distances, magnitudes, figure and composition, but nothing at all concerning their internal composition in respect to the matter from which they are built, their evolutions, their fates, their inhabitants etc. Since, in short, we know of and have no other connection with them but to the eye, then, apart from this, no distinguishing features can be identified and then later arranged, then the natural historian as such must relinquish knowledge and make do with hypotheses.

It was important to draw attention to the relations of our Earth to these bodies, since this shows not only the necessity of this knowledge in order to achieve more exact, genuine, knowledge of so-called natural history, but also the necessity of that science in judging the phenomena [228] of our Earth as a new field of discoveries.

[VI.] In accordance with the above-mentioned axiom of order in Natural History, there should now follow a closer examination of the phenomena produced on the Earth as a whole, since these are for the most part to be viewed, first, as consequences and as always co-existent celestial phenomena and, conversely, the phenomena specific to our terrestrial surface, such as these now are, are to be viewed in relation to their following upon and coexisting with one another.

The history of the phenomena yielded by our Earth as a whole must, according to the concept of natural history, address not only the question of their present state, but also that of the states preceding and perhaps succeeding the present one – thus, how it *is*, how it *was*, and how it *will be*. Since all changes, actions in the world, no matter how slight they may be, are consequents, just as the world itself is the consequence of an antecedent, the present state of our Earth as a whole also had an antecedent state. From the present state we therefore infer in accordance with those phenomena that are forthcoming, since they are not consequences of the present correlation of the Earth with the Heavens or the phenomena of the Earth; we infer a change in the phenomena of our Earth, which [229] do not take place in accordance with the laws that now obtain or are perceived. That Spring arises at a particular time and displays the accompanying phenomenon of trees breaking into leaf is congruent with the current correlation of things. But when sea creatures are permeated with mineral masses and are gathered in large numbers on dry land and are found intact at great depths, and this, throughout almost the majority of known dry land, this is a phenomenon that does not follow from the current course of things, as far as its laws are known to us. One could even say that it is the consequence of the observed decrease or rather retreat of the sea, and that, just as the sea daily absorbs a part of dry land into itself, so the sea creatures that had died out were covered up to the height corresponding to the depth at which we found them. Other circumstances, however, are tied up with this, like the simultaneous presence of the fossilized bones of land animals and other phenomena rather suggest a suddenly precipitated change in the relation.

Animal remains are similarly encountered that are particular to the warmer zones in cold northern lands, elephant bones in Siberia and bones and teeth from fur seals and Rhinoceros in the temperate regions of the Scharzfeld caves in the Harz [mountains], and indeed in large numbers. Again, this phenomenon neither is in accordance with the current state [230] of things nor does it follow from it. We could even say that the angle made by the ecliptic to the equator, which is [now], according to the accepted law of alteration, 23° etc.,[9] was equivalent to a right angle several millennia ago, assuming that, in the course of this [development], the Earth's axis has not changed and, in consequence, at that time for its inhabitants, the Earth reached its zenith and the different zones did not arise at all. And perhaps, in consequence. those fossilised animal remains were able to survive for so long, since soft parts (cheese mites) under certain circumstances fend off destruction for centuries and lead from there to here. Here too, however, difficulties become apparent – in short, we are arguing for a change in the interrelation of phenomena on our Earth. Now this was either effected, gradually or suddenly, by forces that still exist, yet that only produced noticeable effects after a long time; or by forces not currently working directly on them. In favour of the latter case, earthquakes, sudden crevasses and eruptions etc. serve as examples. The alterations themselves can occur with respect to the present properties of our globe either without a change in the universal properties of the terrestrial sphere or with a change in them, or also by forces working from outside, [e.g.] from the heavens, such as comets etc.

Not only the history of the changing [231] surface, but also the history of the state that preceded this change would ideally be an object of Natural History. Since, however, humanity has obscured, as it were, through this first alteration, the facts from which he could in the past have drawn conclusions; since therefore, in this case, simply from what has hypothetically been altered or from what he considers the necessary facts for an alteration, [humanity] had to reason from these changes to the primitive state, *x*; since it is so difficult to ascertain the changed from the unchanging among the effects of daily confusion and the ensuing destruction of the body of our Earth; since the laws of appearance are not so well known to us that we may deduce the past; since here many of the appearances that would be drawn on as proof are partly not established facts and partly unproven; [from all this it follows that] these laws play no role here, since only that to which the facts guide [humanity] on the surface of the Earth belongs to this science.

Inquiring into the original formation of our terrestrial body, perhaps devastated by fire or water and perhaps previously part of another planetary body, has no place here, nor in education, since it can only imagine humanity according to the laws of mechanics. – Compare Kant, *Berl[iner] Monatsschr[ift]*.[10]

[232] Since all these inquiries arise in geography [*Erdbeschreibung*] and since what properly belongs to Natural History can only appear where previously things were to be described as they were (since something can only be said concerning the kind of change when more is known about change), it would have sufficed here to have indicated how important this inquiry into special Natural History is, [for] the classification of animals, vegetables, and mineral formations depends on it.

[VII.] The history of the currently available phenomena of the Earth as a whole, the consideration of its figure, magnitude, motion and its materials (concerning which nothing in general can be said other than that it is unique of its kind and that it is similar to other non-luminous heavenly bodies) does not belong here for the above noted reasons, just as the laws in accordance with which future changes may be made belong to the physicist.

[VIII.] Following the explanation of the influence of the universal properties of the Earth on particular phenomena arising on the surface of our Earth, it is now necessary to touch on the phenomena themselves and to indicate the position in which the object of special Natural History finds itself.
[233] A fleeting glance over our habitat immediately makes us aware of three distinct strata into which the planet is divided: the location of these principal strata accords with the difference in their specific gravity – first the aerial stratum, then the fluid and then the solid.

[IX.] The character of the first is elasticity; air, light and the parts absorbed into them from the other strata provide their particles. A number of phenomena, which can be classified according to the diversity of these universal constituents, become evident here: the first constituent is the medium of sound; luminosity, the fluid, the earths and motions are the objects of a specific science, meteorology. Without this stratum, humanity would lose its sense of hearing, without it the plant would not grow but would be destroyed; without it only a very few animals would be living, and a great number of changes within our terrestrial body, a great number of mineral bodies (birds, flight, insects, infusoria) would not be without it.

[X.] The second stratum, the character of which is fluidity, consists in water, and is what can be bound with solid bodies and thereby preserve fluidity (solution). Movements and mixtures here also generate a great quantity of phenomena. Again, the stratum itself has the greatest influence on the lives of plants and animals, [234] and on the modification of minerals. The sense of taste would presumably be lost without this stratum. It is the habitat of the majority of animals.

[XI.] The three strata ... [*The manuscript breaks off here.*]

Translated by Iain Hamilton Grant

Translator's Notes

1 The following incomplete manuscript is tentatively attributed this title (and also the title: 'Knowledge of Nature') in the *GS*, 211–34 – which this translation follows. Square brackets refer to the page numbers in the *GS* and the one original footnote is an addition by the editor. This manuscript was intended, as will become clear, as a series of lecture notes, and so the reasoning is slightly fragmented and the prose disjointed.

2 Kielmeyer here makes direct use of K. L. Reinhold's term, from his *Versuch einer neuen Theorie des menschlichen Vorstellungsvermögens* (Prague-Jena: Wiedemann u. Mauke, 1789): this faculty or power was to supply the ground for the *Critique of Pure Reason* that Kant, Reinhold argued, had not supplied. This usage suggests Kielmeyer's familiarity with key debates in contemporary philosophy – a view which is confirmed by his 1807 letter to Cuvier (translated below).
3 The sections in square brackets were not completed or are no longer extant.
4 This connection predates and invites comparison with Alexander von Humboldt's foundational *Essay on the Geography of Plants*, trans. S. Romanowski (Chicago: Chicago University Press, 2009).
5 In the paragraph that follows, Kielmeyer uses two different words that mean, approximately, 'species': 'Gattung' and 'Geschlecht'.
6 Didynamia is a Linnaean taxonomic group within the plant kingdom. Didynamous plants have two pairs of stamens, one shorter and one longer pair.
7 Barthold Heinrich Brockes' collected poetry appeared under the title *Earthly Enjoyment of God* from 1738–48.
8 Kielmeyer's point is that, from a standpoint of natural history, facts are *outcomes* rather than starting points – there *last*, not *first*.
9 Kielmeyer is discussing what is called the 'obliquity of the ecliptic', currently measured at 23.4°.
10 Kielmeyer is referring to Kant's 'What is Enlightenment?', first published in the *Berliner Monatsschrift* in December 1784.

Chapter 5

IDEAS FOR A DEVELOPMENTAL HISTORY OF THE EARTH AND ITS ORGANISATIONS: LETTER TO WINDISCHMANN, 1804[1]

C. F. Kielmeyer

[205] As regards the scientific content of your letter, my view of the subjects touched upon is briefly as follows. Certainly, the idea of a closely interconnected *developmental history of the earth and the series of organic bodies in turn*, in which the two should *reciprocally* explicate each other, seems worthy of praise. The reason why I hold this idea to be correct is: because I hold *the force* by which the *series* of organic bodies was once brought forth on our earth, to be *one and the same* as regards its essence and laws, as that force by which even now the series of developmental states is produced in every organic *individual*, which is *akin* to that of the series of organizations. If it is the same force in both cases, and if in the latter case it has an *analogy with magnetism*, as I sought to show in a lecture in 1792, in an analytical way and *starting from experience*, then it must also be assumed in the first case to be analogous to the *magnetism of our earth* and *continuous with it* in some way. Now since the *changes in our earth also affect its magnetism* and are assumed to change in strength, direction etc., then the effects of such a force according to those changes must also have turned out to differ – hence the organizations that are produced, the children of the earth. – In comparison with each other, however, the organizations produced demonstrate a certain regularity of degree in formation and, moreover, a similarity to the developmental states in one individual organization; thus it may also be concluded retrospectively that even the changes that our earth has suffered and namely, that its magnetism [suffers] *with regularity*, were *developmental changes*; and it can therefore be concluded in the end that *the developments of our earth* and those of the series of organic bodies are entirely *continuous with each other* and precisely therefore, their histories must be connected. This, in brief, is the reason, or the reasons, why I hold your idea of such a history to be correct. I believe I have explained these reasons somewhat more clearly than I have here, on p. 38 of the *Speech*[2] with which you are familiar, where what is there simply called *force*, when it appears out of context, is to be interpreted as *magnetism*, and with more clarity in those parts

of my 1792–3 *Zoological Lectures*[3], which have as their object the *theory of developmental phenomena of organizations* and the consideration [207] of the *gradation* and *affinity* of organizations and their causes, as well as the distribution of the animals on our earth and its causes. In short, its object is the predicates attributed to the whole kingdom of organizations as such. You will not likely have seen this in the Ms, since only *a few exemplars* of it exist.

While I indeed understand you according to all this with regard to the idea of a closely-linked developmental history of the earth and its organizations, I do, however, have my doubts regarding the possibility of the execution of such a history, firstly due to time, and, in particular, because each corresponding developmental moment of the earth and, in turn, of the organizations [would need to] be detectable and, as it were, demonstrable in the currently existing series of organizations with any determinacy at all. I doubt the executability, but not the reasons you have provided, because, that is, in all probability, *quite a few intermediary links have been lost from the series of organizations*, which is now *no longer* complete and cannot be seen as the reflection of what was once present. This would practically mean, it seems to me, that a *progression from individually given links* could not be completed, even if it were possible; but it is highly probable that the *series was never continuous* – there are not only *branches* in the network or branching series, but rather *gaps* or *intervals* [208] between one formation and the others, as is visible even now. I surmise this from the stage of formations that every completed plant exhibits in its successively developed parts. Such a plant is a series of similar organizations producing each other and remaining continuous with one another, but no *continuous series* is presented there, precisely because an interval is introduced between stamen and corolla, etc. Likewise for the series of organizations. These intervals seem to act as do distances and intervals that arise when a series of bodies standing in close contact with each other is struck on the first [body], *the last link [or body] is moved farthest* – the more condensed the organization, the greater the gulf by which it is separated from its neighbour and its closest counterpart – thus, *humankind*, as the last link, stands so far removed from all the others. – But the reason why I doubt the execution of such a history is possible is primarily because we *know nothing* about the *development of our earth and its magnetism* and therefore regarding a law of these developments, and just as little concerning a *law to which the gradation of existing* organizations conforms. A comparison of the two and the discovery of a concordance of the two laws is therefore also not yet possible, still less, therefore, is the specification of the *points of coincidence* of the two [209] developments. I formerly clung to some hope of finding a law for the *gradation* in the form and structure of the series of organizations from the aforementioned similar *gradation* evident in the forms of every composite individual plant, but I could advance neither this undertaking nor that regarding the laws of magnetism, and *therefore*, in part, as I have occasionally remarked, the *edition of my inquiries remains undone*, which I hold to be *incomplete* without the foregoing.

Secondly, however, I doubt that it can be executed precisely because I hold the *route* by which the series of organizations have been produced on our earth to be very *different* at *different times*. Many species of organization have probably been produced from each other, just as even now the butterfly is produced from the caterpillar, the flowering parts from the leaf or from the rest of the plant. *At the beginning,* they were *developmental states* and became sanctioned *only later as permanent species*; they are *transformed developmental states. Others*, by contrast, are the *original children* of the life-giving earth. – Perhaps, however, all these *primitive elders* have entirely *died out*. If such a variety is visible now in production, then I no longer see how the *moments of development in the series of organizations* are even to be demonstrable in the moments of terrestrial development.

[210] *Thirdly* and finally, what would make me reluctant to believe that the execution of such a history is possible is the *possibility* that a *regression* [*Zurücksinken*], as Schelling believes, from a better into a lesser form may have taken place; the *possibility* that the organizations could in part have been *alien* arrivals from another planet or *generated* by our earth, as they have an entirely different *connection* with the other planets; finally, the *certainty* that in a *certain respect* not all the changes of our earth are developmental changes. – As concerns the loss of single members from the series, it seems to me, for the above reasons, also on that account not significant, because it appears to me not yet proven. Rather it seems to me, as also Lamarck urges, that everything that has been put forward as such a missing link [*Ausfallen*] can also be explicable as a changed *direction in the forming force* that arose with the *changes* of our earth. [...]

Translated by Iain Hamilton Grant

Translator's Notes

1 This letter was sent to K. J. H. Windischmann on 25 November 1804; it is printed in *GS*, 203–10, where it is attributed the additional title by the editor. This translation follows the *GS* edition of the text and the numbers in square brackets refer to the original. K. J. H. Windischmann (1775–1839) was then Professor of Philosophy and History in Aschaffenburg – a correspondent and sometime disciple of Schellingian philosophy of nature and influenced markedly by Neoplatonic sources, in late 1804 he had just completed his major early work, *Ideen zur Physik*.
2 See p. 42–3.
3 A sketch of these lectures is preserved in *GS*, 13–29.

Chapter 6

ON KANT AND GERMAN PHILOSOPHY OF NATURE: LETTER TO CUVIER, 1807[1]

C. F. Kielmeyer

Tübingen, December 1807

Dear Friend!

I received the confidential letter that you sent me care of Dr. Schnurrer[2] later than you might have expected because the courier lingered in a few places on his journey here and was en route for about three weeks. Owing in part to this circumstance, and in part owing to the content of your letter, which required some reflection, I ask that you excuse my late reply. I freely admit that, had you not asked for a prompt response, you would have been waiting even longer for me to give you a more complete [answer] and more explanation of the reasons for it than has been possible here. What you now receive, as you wished, is a brief exposition of my *opinion* as regards the two questions put to me, more as regards results than reasons. On [236] this subject, I must note that my opinion does not claim to be decisive and should be taken as nothing more than an indication of my personal view.

The first of your questions concerns the assertion that external nature, and consequently facts, may be deduced from *a priori* principles, i.e., such as obtain prior to all experience (not exactly, as you put it, the *abstract* principles that experience may presuppose) – in short, from the nature of our minds.[3] This assertion seems to me to originate in self-deception, but even once this delusion is later recognised as such, it is just pretence maintained by subtle sophistry. The reason for my judgement is, briefly, to be found in the fact that those who claim it *possible* to deduce facts from *a priori* principles and to construct the entirety of nature from our mind, are and must *either* be sheer *Idealists* of the usual sort, who deny the existence of objects outside us and treat all things as a mere representations[4] without really trying to explain the emergence of such representations; *or* they somehow assume objects only to be produced in and by us while at the same time seeking to explain their means of production from the nature of our mind, the original activity of our *I*, and the laws of this activity.[5] Now surely the assumption [237] that there are no objects outside us is permitted here to the extent that we would then, in fact, receive nothing but representations, and anything we know would be of these alone. Yet the first kind of idealist is already contradicted by humanity's first, innate philosophy, the philosophy of language that talks always of suffering and acting, of something objective and subjective, and can never abandon this distinction. Nor have these first type of Idealists ever managed to construct nature, or to prove the nature existing in their representations from the laws of their mind. They formed a sect of innocents or, not ever being a sect, every now and then this arbitrary Idealism is espoused as an opinion, an exercise in scepticism.

Idealists of the second kind, who really are being taken seriously and who are ascendant in Germany in this new era,[6] are contradicted by the fact that, in order to explain the production of the object in us,

1. *either* a stimulus (impulse), by which the activity of our mind, our original *active I*, is determined, gives rise to the positing of an object without further *postulating* a Not-I, that is, an external world of objects;
2. *or*, in order to explain the fertilisation, as it were, of our mind, its incitement to action and the consequent emergence, in the I, of the antithesis between the I and the not-I, subject and [238] object, they *fantasise* that this occurs just as the two perceptibly opposite electricities are generated from an imperceptible, dormant, latent electricity;
3. *or* they leave the emergence of the antithesis of the I and the not-I, subject and object, mind and nature unexplained, positing it as a basic fact insusceptible of further explanation.

In both the first two cases (which in reality are almost indistinguishable, since a stimulus would also be required to generate the two electrical antitheses from their state of insensible indifference), it is clear that the production of objects in our mind that they sought to explain is already presupposed in the assumption of a stimulus acting on it, and is not therefore really explained. In the third case, the assumption of the nonexistence of objects outside us, their being produced in us alone, is entirely *arbitrary* and contains a *petitio principii*. Accordingly, it seems to me that the assertion that an objective world is produced in us by virtue of the spontaneity [*Selbsttätigkeit*] of our I alone remains quite unproven; and, alongside such a world produced in us, were we at the same time to assume a corresponding objective world outside us, with some form of pre-established harmony between them, so that we were no longer merely to say with Leibniz that the soul is a mirror of the universe,[7] but also that the universe is a mirror of the soul, then the first [239] thesis in this case, which institutes the production of an objective world in us through the spontaneity of our I alone, appears as unprovable as before; and for the same reasons, the second thesis of this case – that at the same time there actually exists an objective world outside us – appears superfluous next to the first. However, even if it is accepted that the sheer spontaneity of our I opposes this I to a not-I, that is, an object, neither does this not-I posit nor explain the *quantitative* or the *qualitative* constitution of the objective world. The not-I remains a sterile not-I, and every attempt to conjure the infinite *diversity* of nature from the not-I is futile. The not-I is a metal, a plant, an animal and so on; but metal, plant, or animal and so on are not simply not-I's. Accordingly, it seems to me that these attempts lack both a proof of the principle from which the deduction will be made, and something conditioned by this principle that is to be deduced and by which it is to be understood. Of course, the world appears freely as a simple, ephemeral rainbow in our minds; but the colours of this rainbow are not yet given with the not-I.

All this is not to deny that, as regards the genuinely important purpose of bringing unity to all human knowledge, these attempts to derive it from a single principle and [240] to provide a further exploration and hence a more complete genealogy of our knowledge, along with a genealogical table of nature itself, remain commendable. To attempt something great is already great in itself, and *Plato* and *Leibniz* would be unashamed by many of these attempts, and now and then Kepler himself had more than isolated fragments in view. Nor again is it to be denied that these attempts deserve not merely to be excused, but rather deserve approval in consideration of their courageousness and their purpose of saving man's moral freedom by presenting the objective as the product of his mind. For as long as there exists a determinant external world, proof of the absolute, unconditioned freedom that morality demands, is not possible; but it becomes possible through the hypothesis that the objective world is itself a product of our mind, or that an existent, objective world just exists in harmony and correspondence with the one produced by us.

Finally, it is not to be denied that, in our knowledge of nature, we must accept that a great deal of the subjective and the formal must be acknowledged as coming, not from the external objects themselves, but rather, in the first place, from what is brought to it by virtue of the nature of our mind. For the assumption that our mind acts merely passively towards external nature, like a device that permits recording, a mirror that merely reflects, is, because of its *activity* [241] and its *forces*, quite unacceptable, and even less acceptable than it is acceptable to assume that when, in chemistry, two materials reciprocally effect one another, one is passive and the other active. But if our mind, in contemplating nature, does not act merely passively, then the results of this contemplation too – the representations arising in us – are simply expressions of the relations of our active minds and the object to one another, simply composites of the activities of our minds and the objects. And since we really know absolutely nothing of the objects themselves as such, but only of their relations to us, then objects would absolutely be = x for us. Indeed, if in these relations our mind is more active than objects are, which are reasonably assumed to be more passive, so the results of this composite of the two must also be held to be more subjective and brought to the object by us, just as, in the mixing of two materials, one of which is more active and the other inert, the results of the mixture must bear more of an active character in itself. More remarkably still, this is roughly what occurs when an active organism absorbs such materials as nutrients and so combines them with itself, but [still] expresses only its nature without taking on the nature of those nutrients or being neutralized by them. Now, were our mind, during the intellect's activity of contemplating objects, to behave in same the way [242] as I have just figuratively described the organism as doing, then clearly we would know absolutely nothing about the object, and the results of these contemplations would be entirely subjective, leaving only a small step to full-blown Idealism (all our representations would be an assimilation of the external world with and into us). With our natural cognition being this mixture of the subjective and the objective, it would then be a matter of carrying out a chemical dissolution of the two, as it were, and assigning one of them, the subjective (as existing *a priori* before all experience and even conditioning the possibility of experience itself, albeit only discerned, and perhaps only discernible in experience), perhaps to a specific discipline, the metaphysics of natural knowing [*Naturwissens*], reserving the product of both the subjective and the objective to the empirical cognition of nature. Now for the theoretically minded in Germany, the separation of the subjective from the objective in our knowing provided the initial impetus towards new philosophical systems, after David Hume's sceptical attempts in England [*sic*], which denied any causal connection between phenomena and explained them as merely subjective and, on that account, *not necessary*, introduced this distinction to them. Kant was the first in Germany to recognize this distinction as a necessary one and to investigate it properly. For reasons [243] I will not go into, he explained all knowledge from spatial and temporal relations alone, and consequently explained all mathematical knowledge, the causal connections in which we posit phenomena, etc., as subjective and formal, as something present within us that we have brought with us, and necessarily brought with us, to the contemplation of appearances, and by which, as it were, we precipitate them: [knowledge is accordingly] something through which appearances first come to be manifest for us and come to be appearances, and so which loses its meaning for us beyond appearances and admits of no application to the supersensible. By contrast, Kant considered that which fills space and time for us and thereby delimits itself – in short, the manifold or what appears – to be the operation [*Wirkung*] of some unknown objectivity. He thus avoided the obstacle confronting those Idealists who wished to construct nature from and in itself, namely, the obstacle of explaining the very multiplicity of nature. *Kant's* attempt is astute, and recommends itself in that, by this means, the necessary, universal and the undisputed in our knowing is left subjectively in our mind, while contrariwise, the contingent and particular should be ascribed to an objective nature in itself outside us. This attempt also restricts all *a priori* natural cognition to the formal and its extension over the material and the phenomenological is left to experience. As much, [244] however, as this Kantian experiment recommends itself, it is immediately obvious that it too has significant weaknesses. For to mention but a few, Kant maintains that *matter is the moveable in space, and that this mobility is brought about by a force striving to fill and to limit space, by repulsion and attraction, or better, by expansive and attractive force.*[8] He thus assumes that the essence of matter consists simply in the conflict and unity of what he holds to be two original forces that cannot be further explained. He also therefore assumes that the forces by which matter in general *operates* are also those through which matter is or exists. Yet, were the *essence* of matter in general posited as consisting in the sheer coexistence of these two forces, it would follow from this that the *difference* of matters would depend solely on different quantitative proportions of the two forces; gold and limestone would differ only in the quantitative proportions of the expansive and attractive forces in them. Now, since the quantitative proportions are determinable *a priori*, so one should think that the whole range of varieties of matter on our earth could easily be worked out and determined *a priori* from the concept of matter in general. *Kant*, in order to explain the diversity of nature – that obstacle to the Idealists which, by means of separating the subjective and the objective in our knowing, he seemed [245] fortunately to have avoided – would unexpectedly face this same obstacle again; and we could equally ask him to explain the natural variation as a consequence of his principles and to construct these principles *a priori*. But, apart from the fact that Kant neither achieved this nor wanted to (notwithstanding that he should have), he also still owes us a proof that all qualitative differences amongst materials are simply and immediately differences in the quantitative proportions of the attractive and repulsive forces. I would be pleased to see someone undertake this proof[9] and explain material qualities from these two forces without the intervention of some *tertium*, be it God, atoms, or some third force. Other than this objection to a minor part of the Kantian system, which would only remove the effective assumption of such a *tertium*, a further reproach arises against Kant's attempt. Since the subjective in our knowledge of nature must condition the possibility of experience and, according to Kant, is only known and knowable through experience, one does not rightly see how this subjective element could at the same time be present prior to experience, and how there could be any question at all of something present prior to experience. But, even if we were not overly concerned about this, we do not in the least understand how this subjective [element] could be *known about* or *determined* as regards its detailed composition, *apart from* the objective [element] [246] through which alone it is supposed to be knowable. Surely one should think instead that the subjective [element] would always remain unknown in itself, an *x*, precisely as the object in Kant's system remains unknown in itself. Should we further wish to carry out this separation of the subjective from the objective by abstraction, acknowledging as subjective that which presents itself without exception in all *appearances* and without which we would not be able to represent appearances, namely the spatial and the temporal, then it would first have to be shown that precisely the universal (*as discovered through abstraction*) and the necessary as such *must* also be equally *subjective*. For the spatial in appearances, for example, could also be universal and necessary *either* because extension really attaches to objects themselves in general and belongs to their essence, and could not therefore be represented without this *single* characteristic *common to all*; *or* because the spatial (the extended) was a result of the coexistence and interactions of objects with the forces through which they exist. *Or*, finally and most probably, because the spatial would be a result of the objects' operations on the determinate nature of our minds. In this last case, the spatial as such is attached neither to the subject nor to the object, [247] just as little as the tart, astringent, inky taste evinced by iron, as well as almost all acids in a compound, belongs to the iron itself or to the acids themselves, and indeed would universally and necessarily underlie all phenomena. With regard to this latter possible case, the separation of the subjective from the objective in our knowledge of nature in this manner indeed appears to be a completely impossible task, particularly since similar comments could also be made in respect of the representation of time, in respect to the causal connection in which we place appearances, which appear *not at all necessary* to me. If, however, it is also acknowledged that – as Kant wants – *space and time are subjective*, then the manifold, when treated as such and determined according to its specific type, in our representations of objects, is just as universal, indeed as necessary, as spatial, in so far as we cannot represent objects to ourselves without boundedness, but boundedness is only produced by the manifold in space and what is active in it. One would therefore have to explain not only the spatial and temporal, but also the *bounding manifold* (taking this *in abstracto*) as subjective and formal, *contrary* to Kant's view, and explain them on the same grounds as the foregoing. For there would no more be a geometry without prior experience [248] of the manifold than there would be an experience without geometry. All that remains for the objective is the particular kind of manifold. The separation as carried out by Kant between the subjective and the objective therefore seems to me uncertain and unproven, indicating only a possible case, while an entirely definite and certain separation of the subjective and the objective in our knowledge of nature belongs amongst the most impossible of tasks. Given this weakness in Kant's attempt to determine this separation, it was no wonder that, in Germany after Kant, the Idealists arose. They explained not only the formal in our knowledge of nature, but also the material; not only the mathematical but also the physical, subjectively, and viewed nature *either* as a *mere product* of the lawful operations of our minds; *or*, because they would not then be able to explain its manifold, they assumed, simultaneously with the nature produced in us, a harmoniously corresponding objective nature; thus a productive emission or radiation from the world-spirit within us, and a producing world-spirit, whose individualised emissions we would be and who coincided with his own, produced, objective nature, somewhat as I myself put it (if I have not forgotten it) in the enclosed poem – *The Interpretation of Lethe* – on the death of a young friend who [249] studied here.[10] Now it seems to me that the systems of those Idealists, or rather Real-Idealists[11] that seem already to be partly stated in essence already in the sublime philosophies and religious systems of the Indians and the Persians, are more consistent, although the above-noted objections also hold against them. That's regarding your first question.

Concerning your second question, namely, what sort of success has been achieved, *in my view*, by those whose efforts have been to deduce facts *a priori* in accordance with those systems, this question has in large part been answered by my earlier remarks on the first question, in that, against those systems, I have claimed that, up till now, not one of them has explained the multiplicity of nature and been able to construct it from our *I*. They have, nevertheless, had some success. *First*, that is, we have learnt through them that in our knowledge of nature there really is much that is, could be or must be subjective, although it has not definitively been shown what this subjectivity and the objective remainder somehow consists in. *Secondly*, Kant's idea that the forces by which matters act are also those by which matters exist and accord with their essence, has opened a dynamic view of nature, common since Epicurus and Lucretius, but particularly more widely developed and supported by means of geometry by Lesage [250] in Geneva. Although this has not supplanted mechanical and atomistic methods of explanation, it has rather overshadowed them. For, although Kant's idea has not yet been proven entirely accurate and, if the concept of matter is to be considered exhaustive – as noted above – it too gives rise to unacceptable consequences, the dynamical view of matter retains distinct advantages in explaining the chemical phenomena, the solution and so on of matter as well as the phenomena of the transformation of matter in organic bodies, and so on. *Thirdly*, we have become more aware of the dominance, throughout nature, of the antithesis of the two forces by which phenomena are effected; more aware of the dualism of those forces to the extent that it is *universal* than was hitherto the case. It ought also to be noted, however, that this dualism was already previously acknowledged in particular cases, transposed by analogy onto other cases, and inductively universalised. It cannot therefore be considered simply a consequence and a discovery produced from one of these metaphysical systems. I could show this particularly in regard to the organism and physiological phenomena, for – already in the years 1790 to 1792, well before the philosophical systems after Kant were being discussed – starting not from any philosophical system [251] but merely following experience and analogy, I sought to prove, and was perhaps the first to credibly do so, that the vital operations in organic bodies in general and the developmental effects, especially as manifest in rigid and fluid parts, were to be ascribed to a force working dually towards its poles and acting as magnetism and electricity do – its laws for the most part [being] in harmony with the laws of these forces. In fact, this attempt of mine, which I considered new at that time, did not remain unfamiliar to the founders of those post-Kantian philosophical systems,[12] so that some a posteriori influences might then be discovered in those *a priori* systems. *Fourthly*, through the newer philosophical systems, we have also become more accustomed than hitherto to consider nature in general and as a whole as a living organism in all its operations, and single organizations as individual representatives of nature at large. We have become, that is, more accustomed to an idea that was nevertheless already present in the old antithesis of microcosm and macrocosm, organism and universe. *Fifthly* and finally, it cannot be denied that from the authors of these new philosophical systems, some individual general remarks and some bold, large-scale views have been expressed regarding specific natural objects, e.g. light.

We would be just as wrong, however, to ascribe [252] these remarks and perspectives to their philosophical systems rather than to their intellectual productivity, their fertile imagination and ingenuity, as we would be wrong to derive Buffon's bold and sweeping perspectives from some philosophical system he had adopted, rather than from his productive intellect and his genius. Moreover, such observations and perspectives do not really have *any necessary* connection with those philosophical systems, but are at best *in conformity* with them. Furthermore, in most cases, these perspectives have also been expressed earlier, if only in part. Strictly speaking, therefore, I would restrict the advances made in the natural sciences by these investigations to what is stated in remarks one and two, and I quite agree with you and others that these investigations into our knowledge of nature have done more harm than good, particularly to younger persons in Germany. Some, by the way, remained aware of how to keep themselves free from these systems' influence. I have just read what you wrote concerning your ill-treatment in the *Jena Journal of General Literature*,[13] and I can assure you that this has aroused more irritation against the reviewers, whom I don't know. So much for your questions, which I wanted to be able to answer to your satisfaction.

[253] Perhaps you would be so good as to communicate with [Alexander] von Humboldt, who is now in Paris, and Dr Hochstetter[14] regarding these same subjects and questions and have my judgement amended by them, since I would want the truth. Please give my heartfelt greetings to von Humboldt and, as well as extending him my warmest regards, tell him that I regret the delayed appearance of his book[15] in Germany, of which so far nothing has been seen but the zoological notebooks, due to which I have even refrained from consulting him by letter to inquire about some relevant topics of plant physics. If he could, without detriment, spare one of the seeds he brought back with him, then the local garden would thank him for it. My heartfelt thanks for the *Mémoires* that you sent over by way of Dr Schnurrer and Dr Butte. Perhaps the next time Dr Hochstetter travels here, you could arrange for him to bring with him the last three volumes of your comparative anatomy[16] and the *Ménagerie nationale*,[17] which I have long wanted, and of which I possess only the first volume, then in this respect I will make no further requests and leave it to your discretion. If, with one of my previous letters, you have received a Latin essay that I sent over to you as a keepsake, [254] please acknowledge this in a few words if the opportunity arises in your next letter.

Translated by Iain Hamilton Grant

Translator's Notes

1 The following letter is printed in *GS*, 235–54, where it is there attributed this title by the editor. This translation follows that edition and all numbers in square brackets refer to the pagination in the *GS*. Georges Cuvier (1769–1832): the founder of comparative anatomy and palaeontology had studied in Stuttgart as Kielmeyer's student during the mid-1780s and would elsewhere dub him 'the father of philosophy of nature'. By 1807, he had long established his reputation and, after a series of professorships in natural history, was serving as permanent secretary to the Parisian Academy of Sciences.
2 Friedrich Schnurrer (1784–1833): epidemiologist, then undertaking his studies between Paris and Germany.
3 To understand Kielmeyer's theses concerning the scope of natural history, it is important to note that he does not here present mind as *other than*, but *a variety of* nature.
4 Kielmeyer's term *Vorstellung* is here translated according to its Kantian usage.
5 This is clearly a reference to Fichte.

6 Fichte's philosophy was considered an important impetus in the new theories of physiology issuing from John Brown's *Theory of Medicine*, translated twice into German at the end of the previous century. The circle that congregated around Andreas Röschlaub and his journal in Bamberg was particularly influential in this regard.
7 G. W. Leibniz, *Monadology*, §77: 'the soul (mirror of an indestructible universe)'. It seems unlikely that Kielmeyer is not referring to Schelling here, although the latter had often referred to the mirror-function of spirit – see, e.g. F. W. J. Schelling, *System of Transcendental Idealism*, trans. Peter Heath (Charlottesville: University of Virginia Press, 1978), 92.
8 A reference to the Dynamics chapter of Kant, *Metaphysical Foundations of Natural Science*.
9 As Kielmeyer well knew, this was a project that his former student, A. K. A. Eschenmayer had been undertaking in the years 1797–1801.
10 The poem is reproduced in *GS*, 255–6.
11 A term Schelling often used to describe his project around 1801.
12 Likely a reference to Schelling's praise for Kielmeyer in *On the World-Soul*, in particular.
13 *Jenaische allgemeine Literaturzeitung*
14 K. W. Hochstetter (1781–1811): in 1807, a researcher in Paris, who would go on to become Professor of Medicine at the Academy of Berne from 1810.
15 Alexander von Humboldt's *Observations on Zoology and Comparative Anatomy* was published in 1807–9.
16 Volumes 3, 4 and 5 of Cuvier's five-volume *Leçons d'anatomie comparée*, published between 1800 and 1805.
17 The first volume of Cuvier's *Le Ménagerie du Muséum nationale d'histoire naturelle* (Paris: Miger et Renouard, 1804).

Part III

INTERPRETATIONS

Chapter 7

FORCE AND LAW IN KIELMEYER'S 1793 SPEECH

Andrew Cooper

Carl Friedrich Kielmeyer is an enigmatic figure in the history of science. On 11 February 1793, Kielmeyer gave a speech at the Hohen Karlsschule to celebrate the sixty-fifth birthday of its founder, Carl Eugen. He honoured the occasion with a condensed summary of several years of research in physiology, zoology and natural history. While the lecture offers an extremely difficult, even baffling text to the contemporary reader, it was snapped up by a local publisher and released the same year under the title, *On the Relations of Organic Forces*. The publication was an immense success, and a striking number of Kielmeyer's peers pronounced the speech as a generation-defining achievement that consolidated a range of pressing issues in philosophy, natural history and physiology into a single programme of research. Friedrich Schelling proclaimed it as 'a speech from which in the coming era an entirely new epoch of natural history is expected.'[1] After reading the speech, Alexander von Humboldt dedicated his new research in comparative anatomy to Kielmeyer, announcing him 'the greatest physiologist' of the age.[2] Johann Wolfgang von Goethe cited key passages from the speech in his call for a new science of comparative anatomy.[3] In short, the publication of his speech given at the Karlsschule thrust Kielmeyer into the limelight as a generational '*phenomenon*',[4] leading many to see him as the *Wegbereiter* for a new movement in natural philosophy.[5] After 1793, however, Kielmeyer published virtually nothing. While he continued to develop the speech's methodology in the classroom and through extensive personal correspondence, he never settled on a finished programme of research.[6]

Scholars are in no doubt that Kielmeyer's 1793 speech marks a defining moment in the study of living nature. Yet there is no consensus on what it actually achieved. Edmund Lippmann claims that Kielmeyer advances a vitalist heuristic by which the physiologist can discern organic laws on the 'belief' that the organic sphere is governed by an original, organising force.[7] For Timothy Lenoir, Kielmeyer offers a prolegomena for 'vital materialism', a new programme of research that 'unified mechanistic and teleological principles of explanation.'[8] James Larson goes a step further, arguing that Kielmeyer opens a new system programme for the investigation of living nature in which 'vital forces become causal agents.'[9] On

Larson's reading, Kielmeyer personifies the living forces, thereby enabling the naturalist to examine the manifold of appearances as the result of an action. Peter Hanns Reill consolidates Larson's reading, claiming that Kielmeyer advances a metaphysical form of vitalism.[10]

In this chapter, I propose a different interpretation of the 1793 speech in which Kielmeyer neither unifies mechanism and teleology nor advances a vitalistic account of living phenomena. Alternatively, I argue that he outlines a new standpoint from which the naturalist can examine morphological change as the dynamic interaction of five life forces, thereby *avoiding* any recourse to agential language or speculation about the vital properties of matter.[11] This new standpoint, I suggest, extends the analogical interpretation of Newtonianism developed by Haller, Herder, Blumenbach and Kant, which enabled the naturalist to organise the manifold of living nature according to living forces while remaining indifferent in regard to their cause.[12] Yet, in contrast to his predecessors, Kielmeyer decouples the analogical method from its Newtonian origin. Kielmeyer's speech, I contend, subtly and yet decisively transforms analogical physiology into an independent science of affinity.

This paper is divided into four sections. In the first section, I define the problem confronting eighteenth-century physiologists as *the problem of the manifold*: how the relations between living beings can be determined within the parameters of Newtonian inquiry. In the second section, I examine the use of Newtonian analogies by Haller, Herder, Blumenbach and Kant as responses to this problem. In the final two sections, I turn to Kielmeyer's unique response in the 1793 speech. His method, I propose, reformulates what he terms the 'two chapters' of Kant's *Critique of Pure Reason*, the 'chapter on experience' (the Transcendental Analytic) and the 'chapter on reason' (the Transcendental Dialectic). First, I argue that Kielmeyer extends Kant's doctrine of space and time as forms of intuition to include organic structure. Second, I evaluate the implications of this move for the systematic reconstruction of the manifold according to the demands of reason. I conclude with several remarks on the significance of the 1793 speech for the generation of naturalists following Blumenbach and Kant.

The Problem of the Manifold

Kielmeyer's lecture proposes a ground-breaking solution to the problem of the manifold. During the seventeenth and eighteenth centuries, the integration of experimental practices into philosophy introduced a new standard of scientific explanation based on the physical connections between particles. While this standard led to unprecedented gains in celestial mechanics, and then in particle physics, it simultaneously left the connections between *specific* natural products underdetermined, such as the elements studied in alchemy and the species studied in natural history. In *Origins of Forms and Qualities* (1633), Robert Boyle presented a manifesto for experimental philosophy that rejected the natural system of elements and introduced instead a corpuscular theory of matter. Corpuscularism is a form of monism about the corporeal in which matter is understood as a homogenous, physical substance. Boyle elaborates the consequences of corpuscular philosophy for the natural system of classification as follows:

> Men having taken notice that certain conspicuous accidents were to be found associated in some bodies, they did for conveniency, and for the more expeditious expression of their conceptions, agree to distinguish them into several sorts, which they call genders or species ... as, observing many bodies to agree in being fusible, makeable, heavy, and the like, they gave to that sort of body the name of metal.[13]

On the assumption that qualitative determinations can be reduced to physical connections, Boyle rejects the use of essential qualities to classify natural kinds within the natural system. In *An Essay Concerning Human Understanding* (1689), John Locke extends Boyle's denial to plants and animals. Locke contends that corpuscular philosophy demonstrates that the association of certain individuals within a kind is not based on real essences but rather on the naturalist's interest to gather natural diversity into relations of affinity. He redefined the qualities of an object as ideas in the mind of a perceiving subject, which do not necessarily track natural divisions. 'So uncertain are the Boundaries of *Species* of Animals to us', Locke exclaimed, for we 'have no other Measures than the complex *Ideas* of our own collecting.'[14]

The problem of the manifold is that the affinities we discover in nature are *subjective*. Classificatory marks are not based on the objects themselves, but rather on the connections made between ideas in our minds. While this entails that empirical concepts are artificial, it does not, however, result in full-blown nominalism. Boyle and Locke deny epistemic access to real essences, for ideas do not *necessarily* track real classificatory divisions. Their scepticism is thus generated by a commitment to metaphysical realism, for it is generated by the assumptions that there are natural boundaries to be found. The epistemic limits they place on classification simultaneously open the question of how the boundaries in nature might be reimagined as the result of a natural process and discerned via an experimental method.

Analogical Newtonianism

Isaac Newton altered the course of experimental philosophy by demonstrating how a property that is, at first, merely subjective can be vindicated as a genuine property of an object. In his celestial mechanics, gravity is initially proposed as a hypothetical force to unify the manifold of planetary phenomena. Through the application of mathematics, it is then established as universal and necessary for all planetary movement. The question opened by his work, especially by the ambiguous queries found in later editions of *Principia* and *Opticks*, is whether such a method could be extended to phenomena more specific than matter, such as the bonds between chemical compounds or physiological movement. No one was more successful in providing a positive answer than Göttingen physiologist, Albrecht von Haller. At the close of the eighteenth century, Christoph Heinrich Pfaff commended Haller's achievement with words that foreshadow Schelling's praise of Kielmeyer. Haller's extension of Newton's method to the examination of physiological properties, Pfaff states, initiated a new era 'in the history of the cultivation of physiology.'[15] In this section, I examine Haller's physiology as a ground-breaking response to the problem of the manifold. In the following sections, I argue that Kielmeyer's 1793 speech both extends and surpasses Haller's response.

In a series of two papers presented to the Royal Society of Sciences at Göttingen in 1752, Haller proposed an experimental method for the classification of muscle fibres according to their physiological properties. At first glance, the idea of physiological properties seems incompatible with Newtonianism, understood as the reduction of phenomenal characteristics to the interaction of homogenous particles. However, Haller interpreted Newton's method as a model for investigating movement according to inner properties, without requiring one to speculate about hidden causes. If the Newtonian method is indifferent to hidden causes, then it would not overreach the boundaries of experimental inquiry to apply it to phenomena that do not seem reducible to matter in motion, such as the fibrous matter of muscles. To classify the various kinds of fibres found in animal bodies, Haller proposes two distinct properties, irritability (*Reizbarkeit*) and sensibility (*Empfindung*). These properties do not cause physiological effects, just as gravity does not cause objects to fall. Rather, they stand as methodological postulates by which the naturalist can classify fibres according to their effects. Haller proposes that a fibre is irritable if a contraction can be observed without the organism being sensible to it.[16] It is sensible, in contrast, if stimulation causes the organism to turn the sensation into an image. In the opening line of his textbook on physiology, *Elementa physiologiæ*, Haller identifies the methodological basis of physiology through an analogy with geometry: '*Fibra enim physiologo id est, quod linea geometræ*', the fibre is to physiology what the line is to geometry.[17] The analogy between geometry and physiology enables the naturalist to search for the qualities of known points in a single space based on their relations, and to discover new points within the nexus of connections. The result is an objective procedure by which muscles and nerves can be classified according to their effects.

While most of his successors found the two properties of muscle fibres too simplistic to account for the movement of organic bodies, Haller's investigation opened a new era of physiology by providing a model for extending Newtonian inquiry to living phenomena. In *Ideas for a Philosophy of Human History* (1784), Herder employed the Newtonian analogy to propose three faculties, adding elasticity to Haller's two life forces.[18] Herder's goal, however, was not simply to classify muscle fibres, but to organise the entire manifold of living phenomena under a principle. Herder saw that Haller's method did not acknowledge a difference between the universal properties of matter and the physiological properties of fibrous matter, thereby leaving the question of reduction unanswered. While the properties of matter are universal and reactive, physiological properties are specific and responsive. Herder's solution is to separate mechanical and organic inquiry by identifying two distinct analogies, both of which are grounded in a single, unifying principle. The result is that physiology stands on equal footing with physics: *'where effect is, there must be a force; where new life is, a principle of new life must exist.'*[19] This is to say that the representation of an object as living carries with it, by power of analogy, a transcendental principle of lawfulness. Just as we have warrant to assume that in physical nature every effect is caused by a force, we have equal warrant to assume that living nature is the expression of an organic principle. Herder claimed that his paired analogies can classify living nature under a unifying principle, just as Newton's presumption of a single force that unifies the solar system enabled the discovery of gravity.

Herder's organic analogy received mixed reviews. The idea of a unifying principle seemed to imply a non-Newtonian, teleological power capable of realizing its intentions in nature. Blumenbach outlined an alternative analogy that, in his view at least, solved the Hallerian problem while remaining within the Newtonian frame. He defined the field of physiology according to five hierarchically arranged forces, with Haller's irritability and sensibility at the most general level, and the more specific forces of conception (*Zeugung*), nourishment (*Ernährung*) and reproduction (*Reproduction*) beneath them.[20] Following Herder, Blumenbach's five *Lebenskräfte* are discovered through an analogical procedure by which the naturalist proposes an unknown force as the ground of known effects. Yet Blumenbach denied that the higher forces are governed by a unifying principle. Rather, he identified a further *Lebenskraft* that is limited to organic bodies:

> in the prior brute unformed generative matter of organised beings, before it has arrived at its maturity and determinate place, a particular and enduring active drive [*thätiger Trieb*] begins to take on its determinate shape, which it maintains for its whole life and, if maimed, is able to restore.[21]

Blumenbach names this drive the *Bildungstrieb*, which he presents as the *Lebenskraft* responsible for the form of organised beings. While the idea of a formative drive seems at odds with Newton's mechanical conception of force, Blumenbach insists that the word '*Bildungstrieb*' functions in natural history just like the word 'gravity' in Newtonian mechanics, for it serves 'no more and no less than to signify a force whose constant effect is recognised from experience, and whose *cause*, like the causes of the aforementioned widely recognised natural forces, is for us an *qualitas occulta*.'[22] Like gravity, the *Bildungstrieb* cannot be reduced to more basic powers, for it serves to render intelligible the movement of the parts within a system that cannot be otherwise explained. To examine organised beings as the effect of natural forces is to enable the naturalist 'to give closer determination to their effects and bring them under general laws.'[23]

Kant was impressed by Blumenbach's account of the *Bildungstrieb*, for it demonstrated how teleological properties can be examined alongside mechanical properties of living beings.[24] He was not, however, convinced that the *Bildungstrieb* could be discerned via an analogy with Newtonian forces. In his critical philosophy, Kant aimed to vindicate the use of analogical reflection by separating two uses of analogy. To ensure that analogy is not merely a subjective projection of what reason would like to find in nature, Kant anchors Newtonian inquiry within a conception of the manifold that is not a mere chaos of particulars which must then be reconstructed in the mind according to affinities between ideas. In Kant's epistemology, the manifold is spontaneously presented in spatial and temporal form and schematized with a causal structure. Kant contests that the manifold qualifies as cognition by going through three syntheses: modifications are apprehended in the mind in intuition, reproduced in the imagination and finally recognized in a concept.[25] In the Analogies of Experience, Kant argues that we search for causes only when the connection between phenomena is presented *a priori*. The category of causation provides a transcendental rule connecting two events that are sequential in time (X is followed by Y) as a causal connection, allowing one on appearance of Y_1 to hypothesize X_1 as its cause. The determination of the manifold does not tell us what the cause is, but makes it the case *that* there is a cause to be found.

The upshot of Kant's epistemology is that the causal structure of living beings, by which the parts not only cause the whole (as an aggregate of rock causes a mountain) but the whole also causes the parts (as a living being repairs itself according to an idea of the whole), lies beyond the causal structure of cognition. The capacity of a living being to repair itself, for example, bears a different temporal structure to the linear sequence of appearances *in us*, which is either adjacent in space or sequential in time. Kant's Third Analogy identified the necessary community of parts as a *compostium reale*, which means that each part must be considered within a single causal sequence.[26] Yet the relation between the community and the parts is not causal, for composition is predicated on causal relations between *the parts*. In contrast, the causality of a living natural product moves from parts to whole *and* from whole to parts. Organic movement, including repair and morphological change, is the function of a causality that moves from whole to parts. Even if repair could be explained mechanically, the *possibility* of repair cannot lie in nature understood as mechanism.

In the Transcendental Dialectic, Kant states that when cognition leaves the manifold underdetermined, reason proposes certain ideas that never appear in experience but nevertheless enable the naturalist to unify a particular domain of inquiry, such as the pure elements in chemistry or lines of descent in natural history.[27] On the assumption that specific forces give rise to localized effects, reason aims to unify the manifold by reconstructing it within a projected system. Yet, in contrast to the constitutive ordering of the categories, the ideas of reason do not determine the manifold. They simply guide our reflection, meaning that experimental sciences such as chemistry and natural history cannot determine hypothetical properties as properties of the object; they carry merely subjective necessity. In the third *Critique*, Kant argues that the variation of living beings must be considered as a dynamic interaction between external forces, which act on living beings, and inner purposes, by which the organism gives shape to itself. Yet, inner purposes are not examined by an analogy with the causality of the understanding; they are examined by an analogy with *our own purposiveness* as rational agents, by which a representation of the whole determines our parts.[28] Thus, the *Bildungstrieb* cannot be a law of nature. It is a subjective law for unifying the manifold, and cannot be reduced to the mechanical properties of matter.[29]

Space, Time and the Manifold

Early in his career, Kielmeyer began work on a new conception of natural history that built on Kant's distinction between inner and outer nature. Yet he questioned Kant's twofold conception of analogy, for it denies the investigation of living nature as a matter of cognition. In the following two sections, I suggest that Kielmeyer's conception of natural history transforms Kant's transcendental account of nature in two important ways. First, it reframes Kant's account of space and time such that they include the spatial and temporal form of organised nature. Second, it derives the laws that regulate the distribution of vital functions in the animal kingdom. Together, these modifications enable the naturalist to systematize the manifold according to general laws, thereby opening the prospect of a science of affinity. In this section, I focus on Kielmeyer's account of time and space, beginning with his early writings and then turning to the introduction to the 1793 speech.

In an unpublished paper from 1790, 'On Natural History', Kielmeyer presents the problem of the manifold as a consequence of the finite conditions of human knowledge. In keeping with Kant's epistemology, he affirms the discursive nature of human cognition. Thought does not produce objects by virtue of thinking them. Rather, it gives spatial and temporal form to an objective order of nature that exists independently of us. The discursive nature of cognition thereby entails that natural divisions are not immediately given. Yet Kielmeyer then provides a definition of nature that modifies the Kantian frame. Nature, he states, consists of 'everything emergent [*alles Entstandene*] or actually appearing to our senses [*unsern Sinnen wirklich Scheinende*] and which is perceived with our outer and inner sense, connected in time and space and apt to follow certain laws' (p. 55). If nature does not simply concern the totality of natural products but also the laws by which they became manifest in time and space, then the idea of a *natural* history 'must . . . address not only the question of their present state, but also that of the states preceding and perhaps succeeding the present one – thus, how it *is*, how it *was*, and how it *will be*' (p. 60). In contrast to many experimental philosophers, who consider the manifold as a collection of facts awaiting systematization, Kielmeyer follows Herder by claiming that natural history is possible only according to a transcendental presupposition that the manifold is the product of a causal history.[30]

Kielmeyer's inclusion of 'everything emergent' in the concept of nature signals his attempt to transpose Kant's regulative account of natural products into a constitutive account of natural processes. Kielmeyer shares with the analogical Newtonians the idea that the *scala naturae* is not an object of knowledge but rather a regulative hypothesis that enables the naturalist to search for the forces that determine the connections in one's model. Yet to give this hypothesis transcendental grounding, he replaces the logical conception of the *scala naturae* with the 'series of organisations', the idea of a dynamic and open system that is in a constant state of growth and change.[31] In his lecture notes for a course on the development of organisations given in 1793/4, Kielmeyer repeats the Kantian refrain that, when it comes to organic change, all we experience are alterations that are contingent in regards to the universal laws of nature. However, he then states that:

> In addition to these irregular changes, we note in all organizations a series of definite successive changes which are the same in the individuals of the same genus, and which remain the same even in altered external circumstances, and in contrast to different circumstances in individuals of a different genera, and, therefore, seems to be independent of external circumstances, and seems to be made up more of its own inner forces.
>
> GS, 107

Kielmeyer insists that the changes that occur in organisations are not constructed by the subject. They are the effects of inner forces, which means that they can be investigated in search for principles of change. Given that the development of organisations is not simply a projection of reason's demand, Kielmeyer concludes that they must be considered according to an alternative conception of nature as the totality of connections of effects in time and space, that is, the series of organisations. The goal of natural history is to grasp the laws that govern the development of the series, not simply the regulative principles that govern *our* attempt to comprehend it.

In the 1793 speech, Kielmeyer states that his goal is to develop a 'physics of the animal kingdom [*Physik der Tierreichs*]' through comparative anatomy. In Kant's view, experimental physics is guided by categorized experience, which provides the universal form of causation (X is the cause of Y). Natural history, in contrast, examines natural changes according to the causal form of organised beings (things of *such and such a kind* do Y when X happens). This causal form is foreign to categorized experience, and is derived analogically by extending the form of rational determination as a guide for reflecting on the manifold of living nature. Thus, to ground a *physics* of the animal kingdom, Kielmeyer modifies Kant's account of space and time as forms of intuition such that the unique causal form of organisations *is* a part of nature.[32]

To achieve this goal, Kielmeyer begins the lecture by questioning the privileged position given to cosmology in eighteenth-century natural science. While naturalists have hitherto contemplated the heavens as the greatest display of nature's order, he invites his audience to consider instead the startling and unparalleled order manifest in the organisations of living nature. Our capacity to identify some objects as organisations, as interdependent clusters of organs that manifest internal unity, does not occur on the level of our reflection *on* the manifold. Rather, it occurs on the primitive level of the schematizing understanding, such that organisations are *given within* the manifold:

If, by the powers of our minds, we separate the phenomena of nature – for us connected in a system by space and time – from their connection, then surely those phenomena that we isolate and subsume under the name 'animate nature' – I mean the organisations of our earth – are the most able to fill us with feelings of nature's greatness of those [phenomena] with which we are closely acquainted. To be sure, no masses, volumes, or distances found here are like those of the skies, by which nature convinces us of its greatness. However, if, when judging the greatness of an object, we can deign to give voice and listen with a little patience to the multiplicity [*Vielheit*], diversity [*Mannigfaltigkeit*] and harmony [*Harmonie*] of effects in a small space and short periods of time, then there are things of another kind that speak to us no less forcefully.

p. 30

Rather than turning to the celestial system as the model by which reason organises the manifold presented in cognition, Kielmeyer invites his audience to begin with the pre-structured multiplicity of organised nature, and to discern the harmony of effects and causes that express another kind of order. First, this draws our attention to the incredible diversity of organisations on the surface of the earth, which is a remarkably small space compared to the planetary system. Second, it draws our attention to the manner in which these things occupy time. The changes undergone by an organisation result in the reciprocal adaptation of all the other organs, thus forming a system that is united such that 'each is reciprocally cause and effect of the other' (p. 30). This same configuration characterizes the organisations within a species, and the organisations within an environmental system, which come together to 'form the life of the great machine of the organic world' (p. 30).

Kielmeyer's mechanical metaphor does not imply that the organic world is reducible to physics, nor that it is the product of a pre-structured plan. Rather, it illustrates the systematic combination of forces.[33] Consider his presentation of the machine of the organic world in 'On Natural History':

Since a machine in the broadest sense is a combination of one or more forces (towards a specific end), by the cooperation of which certain results will be obtained, it is permitted, if the larger results are not obvious to us and to the extent that we note interconnectedness, to represent nature as a great machine whose cogs we grasp according to all the phenomena that arise for us.

p. 55

The mechanical metaphor enables the naturalist to look at the emergent levels of complexity on the assumption that each is determined by lower-level connections, without needing to reflect on what those connections are. The emergent system is a dynamic product of several forces that compensate each other within space and time.

In a letter to Cuvier in 1807, Kielmeyer considers the relation between the qualitative distinctness of matter and quantitative relations of forces. He notes that Kant's empirical concept of matter in *Metaphysical Foundations* aimed to show that qualitative distinctness can be reduced to quantitative relations of forces. The essence of matter for Kant lies simply in the communion of attractive and repulsive forces, through which matter effects and exists. Sceptical of Kant's reduction, Kielmeyer takes the discussion in an alternative direction:

> Were the *essence* of matter in general posited as consisting in the sheer coexistence of these two forces, it would follow from this that the *difference* of matters would depend solely on different quantitative proportions of the two forces; gold and limestone would differ only in the quantitative proportions of the expansive and attractive forces in them. Now, since the quantitative proportions are determinable *a priori*, so one should think that the whole range of varieties of matter on our earth could easily be worked out and determined *a priori* from the concept of matter in general. *Kant*, in order to explain the diversity of nature, that obstacle to the Idealists which, by means of separating the subjective and the objective in our knowing, he seemed fortunately to have avoided, would unexpectedly face this same obstacle again; and we could equally ask him to explain the natural variation as a consequence of his principles and to construct these principles *a priori*.
>
> pp. 72–3

Kielmeyer identifies Kant's empirical concept of matter as a moment of transition between experimental philosophy and a new kind of idealism. Kant identified the ground of objectivity in the subjective conditions of experience, yet he nevertheless attempted to reduce qualitive relations to attractive and repulsive forces. What can be known objectively is simply objects as determined by the categories, while the remainder of the manifold is a matter of subjective reconstruction. For Kielmeyer, Kant's failure to determine the qualitative distinctness of matter *a priori* reveals a gap between the manifold and our cognition of objects:

> If, however, it is also acknowledged that – as Kant wants – *space and time are subjective*, then the manifold, when treated as such and determined according to its specific type, in our representations of objects, is just as universal, indeed as necessary, as spatial, in so far as we cannot represent objects to ourselves without boundedness, but boundedness is only produced by the manifold in space and what is active in it. One would therefore have to explain not only the spatial and temporal, but also the *bounding manifold* (taking this *in abstracto*) as subjective and formal, *contrary* to Kant's view, and explain them on the same grounds as the foregoing. For there would no more be a geometry without prior experience of the manifold than there would be an experience without geometry. All that remains for the objective is the particular kind of manifold.
>
> p. 74

On Kielmeyer's account, Kant failed to recognize that the manifold contributes the limits by which the understanding can then schematize objects. What is given in the manifold is more than spatial and temporal form, but also spatial and temporal *effects* by which objects limit *themselves*. By reducing the manifold to what can be constructed as an object in experience, Kant underdetermines the manifold's significance *for* experience. He does, however, make some headway in his account of reflective judgement in the third *Critique*. Yet, reflective judgement can only reflect on the unschematized manifold via subjectively determined analogies. Thus, to expand the Kantian standard of explanation to include reflection, Kielmeyer does not transform Kant's 'heuristic principle into a constitutive agent', as one commentator puts it.[34] Instead, he locates the reflective operation of judgement prior to its determinative application, thereby opening a way to search for and establish the laws of animate nature.

Force and Law in the 1793 Speech

In the introduction to the 1793 speech, Kielmeyer outlines a new method that opens a possible physics of the animal kingdom. The natural historian begins with the great machine of the organic world, understood as a dynamic equilibrium in which the relation between the various forces of living nature are balanced in time and space. She then searches for the laws of animal physics, which establish the boundaries between species groups. Given the unique manner in which organisations occupy space and time, Kielmeyer identifies three basic tasks for animal physics: it must (a) define 'which forces are united in the greatest number of individuals'; (b) discern 'the proportions [*Verhältniße*] of these forces to each other for different species of organisations, and by what laws do these proportions modify themselves in the series of different organisations'; and (c) determine how these relations 'are grounded in [the forces] as their cause' (i.e. how *b* is grounded in *a*) (p. 31). Let us consider each task in turn.

a. The Forces Unified in the Greatest Number of Individuals

In Kant's account of natural science, the naturalist has warrant to propose hypothetical forces in the analogical form X stands to Y as X stands to Z, for she knows that possible objects stand in the relations determined by the categories: succession, causation and simultaneity.[35] Only by presupposing that physical objects stand within a dynamical system in outer sense can the naturalist reflect on possible grounds for their existence and then propose that the same X causes, say, the orbit of the moon *and* the motion of the sea. To extend this analogical procedure to animate nature, Kielmeyer proposes that organisations are a matter of intuition. Like non-living nature, organisations occupy space as an unfathomable manifold. Yet in contrast to non-living nature, they occupy time as a reciprocal relation of cause and effect. The reciprocity of cause and effect is not added by reflecting on the part-whole structure, which in Kant's account was somehow both available to cognition and yet in want of reflective systematization. Rather, organisation is a temporal relation given in intuition. That the part-whole temporality of organisations is given in intuition enables us to propose various forces that govern manifest effects.

In Kielmeyer's methodology, a force is merely a 'makeshift word' [*Behelfwort*]' that ought to be used only 'for now [*einstweilen*]' (p. 32). In Newtonian science, gravity is initially proposed as a problematic force, for its truth remains to be established. It acquires necessity only when the naturalist establishes that no other candidate force could unify the manifold with such simplicity and completeness. Building on this method, Kielmeyer proposes five organic forces: sensibility (*Sensibilität*), irritability (*Irritabilität*), the force of reproduction (*Reproductionskraft*), the force of secretion (*Secretionskraft*) and the force of propulsion (*Propulsionskraft*) (p. 32). In this list, he replaces Blumenbach's conception and nourishment, which are not really forces but capacities, with propulsion and secretion, which concern movement alone.[36] More significantly, Kielmeyer rejects the idea of an organisational force that guides and directs the expression of the five organic forces. He is concerned only with the relations (*Verhältniße*) of the forces, signalling a much more empiricist programme of research to that proposed by Blumenbach.[37]

b. Relations and their Laws

In contrast to the planetary system, which is constituted by the relation between attraction and repulsion, the great machine of the organic world consists of the relations of the five life forces. It is not fixed, but rather a self-generating system that changes itself in time as natural history. The universal principle that structures animate nature is not an active drive but rather a process of exchange by which each force comes to dominance or is repressed in proportion to the other forces. Kielmeyer defines this process as the 'law of compensation [*Kompensationsgesetz*]'.[38] The law of compensation is not a governing relation, for nothing directs the interplay of the forces. Rather, it captures the self-regulation of the forces within the constraints of the earth. Kielmeyer uses the geometrical metaphor of a parabola to explain the openness of the organic system, which 'never closes in on itself' (p. 30). As Thomas Bach explains, the parabola metaphor captures the individual character of the great machine of the organic world, which is regular and yet undergoes continual alteration.[39] It offers a vision of natural history as a 'self-expanding spiral', to use Marie Heuser-Keßler's phrase, which grows in complexity through the dynamic capacity of organised beings to compensate their effects in relation to other effects.[40] Organic systems are thus irreducible to the Kantian concept of matter, for they are constituted by forces that rise or decline in relation to the other forces.

Kielmeyer's rejection of an organising force means that he does not need to make any ontological claims about levels of reality or kinds of matter. Consider one of his many descriptions of the simultaneous and consecutive changes in each organ within an organisation: 'according to our manner of speaking – each is reciprocally cause and effect of the other' (p. 30). Locutions such as this suggest that Kielmeyer views the language of organisations and their laws as pertaining to a phenomenological level of scientific analysis. He claims that even if this phenomenological level were reducible to efficient causation, it would not hamper his new scheme in the slightest:

suppose that nature had no intent in artificially placing phenomena one after another and next to each other in time, that those effects and consequences were not ends that [nature] wanted to achieve ... we would still nevertheless have to admit that in most cases this chain of cause and effect looks like a chain of means and end to us, and we even find it conducive to our reason to accept such a chain; and we must thereby, at least in the end, admit that nature is no less able here than in the skies of convincing us of the truth of that [the greatness of nature] from which I began [i.e. the unique temporal structure of animate nature].

pp. 30–1

Kielmeyer's counterfactual conditional acknowledges that even if nature has no intrinsic purposes – even if organisation is the emergent structure of mechanical forces – we can nevertheless acknowledge the reciprocal causal structure that permits systematic reconstruction no less than celestial mechanics.[41] The principle of compensation does not make it the case that a teleological cause governs the series of organisations; it simply unites the systematic interconnection of the manifold as a natural process, meaning that the system is not closed and complete but spiralling upward. This is nothing less than a Copernican revolution in physiology, such that the dynamic system of living nature can be reconstructed from the subjective standpoint of the inquirer, who is herself part of the system. From such a standpoint, one does not need to ground organisation in an organising force, either hyperphysical or immanent to organic matter. Organisation is rather the shifting relation of the lower three forces, irritability, sensibility and reproduction, which determines the degree of sophistication.

To exemplify this process, let us consider the relation between sensitivity and reproduction. In Kielmeyer's analysis, sensitivity and reproduction can be understood in a relation of polarity. As the series of organisations diminish in complexity from humans to animals and to vegetables, the diversity of sensations diminishes to the point of uniformity of movement.[42] This observation gives rise to the law, 'the manifoldness of possible sensations in the series of organisations decreases as the fluency and refinement of the remaining sensations increases in a more restricted scope' (p. 35). Alternatively, for plants, where forms of reproduction are at their most various, irritability and sensibility are almost non-existent: 'irritability increases as the speed, frequency and diversity of the same expressions, and the diversity of sensations, decreases' (p. 37). The increase in diversity of reproduction thus occurs in reverse proportion to the diversity of sensations, where in plants it receives maximum diversity. This gives rise to a further law: 'the reproductive force ... increases as the size of the producing individuals, or more universally, the individuals produced as they appear after birth decreases' (p. 39). The discovery of the laws of organisation occur through observing the relation between the forces, such that if one force is less present in an organism, the others tend to be more fully developed. The distribution of the organic forces is regulated by the law of compensation, which states that the sum of all forces must remain constant.

The explanatory role of the law of compensation can also be seen in Kielmeyer's account of the relation between sensitivity and reproduction. For sensitivity: as the variety of sensations diminishes across the system of organisations, the degree of refinement of the remaining senses increases. For reproduction: the number of new organisations brought forth by genetic replication increases as the size and complexity of the generated individuals diminishes. Reproduction permits a second law concerning transformation, which effects the capacity of an organisation to substitute organic reproduction with partial organic regeneration or metamorphosis: 'the less the reproductive force is expressed in many new individuals, the more it is expressed either through metamorphoses that the body withstands, or through unusual artificial reproduction, or both together, or by indeterminate growth, or by greater division in the newly begotten formations' (p. 41).

c. Grounding the Laws

The final task of a physics of the animal kingdom is to determine how relations are grounded in the forces as their cause. Here Kielmeyer steps back from the relations themselves to examine the systematic arrangement of the organic world according to the laws discovered in *b*. From this higher vantage, the law of compensation is not the equilibrium that emerges through the vanishing and emergence of opposed forces; rather, it is the equilibrium that emerges through their repulsion and even destruction. Plants, because they possess little or no sensibility and irritability, 'repel all destructive forces of the animal kingdom with their reproductive force' (p. 43). Worms, which are a class between plants and animals, withstand the animal kingdom by virtue of its 'indestructible irritability and reproductive capacity'. The higher animals, who are the most destructive by virtue of their great mass, are supported by the great number of smaller individuals. The animals with greater irritability and sensibility, and thus who move more swiftly and frequently, are given towards predation. The continued existence of the organic world is thus determined by the balancing (*Ausgleichung*) of the forces, captured by a further law: 'destruction by one force was thus regularly repelled by another force, or limited by a different side of the same one, and thereby the species was at least preserved' (p. 44). The law of compensation thus entails that as one force disappears, another emerges. The relation between parts within a plant or animal, the relation between various species, and relation between living beings and the earth are determined by the balancing of the laws. This is how the relations are grounded in the life forces: the 'vanishing of one [force] can therefore be regarded as the cause of the emergence of the other' (p. 42).

Kielmeyer concludes with a remarkable reflection on the human being as the species that possesses the greatest power of compensation of one organic force for another. Rationality 'turns up' in humanity's organisation due to an 'over-preponderance' of the forces, giving rise to 'the capacity to freely alter (within certain limits) the relation of the other forces that it has in common with other animals' (p. 44). Where humans lack in sight or movement, they can create technologies such as telescopes or domesticate other species such as horses, granting to the human species 'a decisive preponderance over most other animal species and their forces of preservation' (p. 44). The human species is uniquely plastic, for it can freely alter the balance of the forces. 'The human spirit [*Geist*]',

Kielmeyer states, 'displays changed proportions [*Verhältniße*] of the forces united in it', such that 'changes of circumstance present these [changes of forces] with their development' (p. 45). Because the compensation of forces is internal to the human organisation, the capacity is analogous to the self-healing of the polyp, which actively responds to privation by supplementing itself with an alternative force. The character of the human species is 'that capacity of compensation [*jene Erstattungs Fähigkeit*], or in other words, the capacity to flourish in all external circumstances' (p. 46). Wisdom and happiness do not lie in the unnatural attempt to steer fate, Kielmeyer concludes, but rather in the cultivation of the natural capacity through which fate is confronted.

Conclusion

Examining Kielmeyer's lecture as a response to the problem of the manifold clarifies its historical significance. During the eighteenth century, natural historians and physiologists responded to the problem of the manifold by transposing the *systema naturae* into a regulative ideal for searching out nature's order. The analogical interpretation of Newtonianism proposed by Haller, Herder, Blumenbach and Kant enabled each naturalist to develop a system of classification according to life forces. In his 1793 speech, Kielmeyer takes this development to its conclusion – and beyond it – by establishing a system of organised nature as a self-expanding spiral, a developing parabola that manifests its systematic connections as the great machine of organic nature. In Bach's terms, Kielmeyer's speech transforms 'the forces of the soul [*Seelenkräfte*] ... into the forces of bodies [*Körperkräften*]', placing the static Newtonian cosmos within a self-generating causal sequence without the addition of a vital organising force.[43] Thus conceived, membership in a genus such as quadrupeds is not determined by the mere fact of having four legs, which is simply an accidental characteristic; rather, it is determined by a specific relation between sensation and reproduction, such that the former is extremely abundant while the latter is constrained to a single mode of expression. The transformation of the forces of the soul to the forces of bodies is of monumental significance for the development of a unified science of organic life, for it rejects the analogy between mechanical and living nature and opens an independent physics of organisation. Life for Kielmeyer *is* the 'system of effects' and an organisation *is* the 'system of organs' (p. 30).[44] As an open spiral, the *systema naturae* can be understood as continuity in change, unity in development, an ever-growing abundance of living forms within the finite economy of the earth.

Notes

1   F. W. J. Schelling, *Von der Weltseele: Eine Hypothese der höhern Physik zur Eklärung des allgemeinen Organismus* (Hamburg: Perthes, 1798), 298. Research for this chapter was supported by an Early Career Fellowship awarded by the Leverhulme Trust. I would

also like to thank the Niedersächsische Staats- und Universitätsbibliothek Göttingen for their generous support in accessing the source materials.
2 A. von Humboldt and A. Bonpland. *Beobachtungen aus der Zoologie und vergleichenden Anatomie* (Tübingen, 1806–9), v.
3 See, for example, J. W. Goethe, *Erster Entwurf einer allgemeinen Einleitung in die vergleichende Anatomie, ausgehend von der Osteologie*, in *Die Schriften zur Naturwissenschaft* (Leopoldina Ausgabe), ed. D. Kuhn (Weimar: Hermann Böhlau Nachfolger, 1947), I/9: 124.
4 K. T. Kanz, 'Carl Friedrich Kielmeyer (1765-1844) – Leben, Werk, Wirkung: Perspektiven der Forschung und Edition', in K. T. Kanz (ed.), *Philosophy des organischen in der Goethezeit: Studien zu Werk und Wirkung des Naturforschers Carl Friedrich Kielmeyer* (Stuttgart: Steiner, 1994), 14.
5 I. Schumacher, 'Karl Friedrich Kielmeyer, ein Wegbereiter neuer Ideen: Der Einfluß seiner Methode des Vergleichens auf die Biologie der Zeit', in *Medizinhistorisches Journal* 14 (1979); K. T. Kanz, 'Carl Friedrich Kielmeyer (1765-1844): Wegbereiter des Entwicklungsgedankens', in H. Albrecht (ed.), *Schwäbische Forscher und Gelehrte* (Stuttgart: DRW-Verlag, 1992).
6 K. T. Kanz, *Kielmeyer-Bibliographie: Verzeichnis der Literatur von und über den Naturforscher Carl Friedrich Kielmeyer (1765-1844)* (Stuttgart: GNT Verlag, 1991), 12. Kuhn speculates that Kielmeyer has been overlooked as a serious contributor to the development of biology due to his few publications, and points instead to his commanding personality as the source of his influence. See D. Kuhn, 'Uhrwerk oder Organismus: Carl Friedrich Kielmeyers System der organischen Kräfte', in K. T. Kanz (ed.), *Philosophy des organischen in der Goethezeit: Studien zu Werk und Wirkung des Naturforschers Carl Friedrich Kielmeyer* (Stuttgart: Steiner, 1994), 37.
7 E. Lippmann, *Urzeugung und Lebenskraft: Zur Geschichte dieser Probleme von den ältesten Zeiten an bis zu den Angängen des 20. Jahrhundrets* (Berlin: Julius Springer, 1933), 94.
8 T. Lenoir, 'Kant, Von Baer, and Causal-Historical Thinking in Biology', in *Poetics Today* 9.1 (1988), 104. See also T. Lenoir, *The Strategy of Life: Teleology and Mechanics in Nineteenth-Century Biology* (Dordrecht: Reidel, 1982), 44–53.
9 J. Larson, 'Vital Forces: Regulative Principles or Constitutive Agents? A Strategy in German Physiology, 1786-1802' in *Isis* 70.2 (1979), 241.
10 P. Reill, *Vitalizing Nature in the Enlightenment* (Berkeley: University of California Press, 2005), 191–2.
11 See E.-M. Engels, 'Die Lebenskraft – metaphysisches Konstrukt oder methodologisches Instrument? Überlegungen zum Status von Lebenskräften in Biologie und Medizin im Deutschland des 18. Jahrhunderts', in K.T. Kanz (ed.), *Philosophy des organischen in der Goethezeit: Studien zu Werk und Wirkung des Naturforschers Carl Friedrich Kielmeyer* (Stuttgart: Steiner, 1994) and *Die Teleologie des Lebendigen: Kritische Überlegungen ur Neuformulierung des Teleologie-problems in der angloamerikanischen Wissenschaftstheorie* (Berlin: Dunker and Humblot, 1982), 97–8.
12 I draw the term 'analogical Newtonianism' from Charles Wolfe's topography of Newtonianisms in 'On the Role of Newtonian Analogies in Eighteenth-Century Life Science', in Z. Biener and E. Schliesser (eds), *Newton and Empiricism* (Oxford: Oxford University Press, 2014), 223–4.
13 R. Boyle, *Works*, ed. M. Hunter and E. B. Davis (London: Pickering and Chatto, 1999), III 36.
14 J. Locke, *An Essay Concerning Human Understanding* (Oxford: Clarendon, 1975), III vi 27, 290.

15 C. H. Pfaff, *Über tierische Elektricität und Reizbarkeit* (Leipzig: Crusius, 1795), 236.
16 A. von Haller, 'A Dissertation on the Sensible and Irritable Parts of Animals,' ed. O. Temkin in *Bul. Hist. Med.* 4 (1936), 692.
17 A. von Haller, *Elementa physiologiæ corporis humani*, (Lausanne: Marc-Michel Bousquet, 1757), I 2.
18 J. G. Herder, *Ideas*; translated as: *Outlines of a Philosophy of the History of Man*, ed. T. Churchill (London: J. Johnson/L. Hansard, 1803).
19 Ibid., 51. I have modified Churchill's translation of *Kraft* from 'power' to 'force' for consistency.
20 J. F. Blumenbach, *Über den Bildungstrieb* (Göttingen: Dieterich, 1791), 25.
21 Ibid., 24.
22 Ibid., 25-6.
23 Ibid., 26.
24 I. Kant, *Critique of the Power of Judgment*, trans. P. Guyer (Cambridge: Cambridge University Press, 2000), 5: 424.
25 I. Kant, *Critique of Pure Reason*, trans. P. Guyer and A. Wood (Cambridge: Cambridge University Press, 1999), A97.
26 Ibid., A215/B262.
27 Ibid., A644/B672-A645/B673.
28 Kant, *Critique of the Power of Judgment*, 5:375.
29 Ibid., 5:424.
30 For a study of Kielmeyer's relation to Herder, see W. Pross, 'Herders Konzept der organischen Kräfte und die Wirkung der Ideen zur Philosophie der Geschichte der Menschheit *auf Carl Friedrich Kielmeyer*', in K. T. Kanz (ed.), *Philosophy des organischen in der Goethezeit: Studien zu Werk und Wirkung des Naturforschers Carl Friedrich Kielmeyer* (Stuttgart: Steiner, 1994).
31 T. Bach, *Biologie und Philosophie bei Kielmeyer und Schelling* (Stuttgart: Frommann-Holzboog, 2001), 93.
32 M.-L. Heuser-Keßler, 'Raum, Zeit, Kraft und Mannigfaltigkeit. Kant und die Forschungsmethodologie der Physik des Organischen in Kielmeyers "Rede"', in K. T. Kanz (ed.), *Philosophy des organischen in der Goethezeit: Studien zu Werk und Wirkung des Naturforschers Carl Friedrich Kielmeyer* (Stuttgart: Steiner, 1994), 113.
33 Bach, *Biologie und Philosophie*, 137.
34 Larson, 'Vital Forces', 241.
35 Kant, *Critique of Pure Reason*, A177/B220.
36 Gambarotto claims that, in the speech, 'Blumenbach's *Bildungstrieb* is renamed as the reproductive force.' Alternatively, I propose that it is not renamed but replaced. A. Gambarotto, *Vital Forces, Teleology, and Organization: Philosophy of Nature and the Rise of Biology in Germany* (Dordrecht: Springer, 2018), 44.
37 Dougherty argues that Kielmeyer's approach to organic forces is more empirical than Blumenbach, whose *Bildungstrieb* had metaphysical implications Kielmeyer wished to avoid. F. Dougherty, 'Über den Einfluß Johann Friedrich Blumenbachs aufs Kielmeyers feierliche Rede von 1793', in K.T. Kanz (ed.), *Philosophy des organischen in der Goethezeit: Studien zu Werk und Wirkung des Naturforschers Carl Friedrich Kielmeyer* (Stuttgart: Steiner, 1994), 60-5.
38 The law can be stated as follows: 'the more one of these forces on one side is cultivated, the more it is neglected on the other' (p. 41).
39 Bach, *Biologie und Philosophie*, 101.
40 Heuser-Keßler, 'Raum, Zeit', 116.

41 R. J. Richards, *The Romantic Conception of Life: Science and Philosophy in the Age of Goethe* (Chicago: University of Chicago Press, 2002), 242.
42 G. Bersier, 'Visualising Carl Friedrich Kielmeyer's Organic Forces: Goethe's Morphology on the Threshold of Evolution', in *Monatshefte* 97.1 (2005), 22.
43 Bach, *Biologie und Philosophie*, 198.
44 See I. Jahn, *Grundzüge der Biologiegeschichte* (Jena: G. Fischer, 1990), 296; Bach, *Biologie und Philosophie*, 134–5.

Chapter 8

ORGANIC PHYSICS AS A PHENOMENOLOGY OF THE ORGANIC[1]

Thomas Bach

It has been fifty years since Dorothea Kuhn – in her classical study of 'Karl Friedrich Kielmeyer's System of Organic Forces' – posed the question of the key to his conception of nature and explained it in terms of the concepts of 'clockwork' and 'organism'.[2] Kuhn argues that Kielmeyer started out from the idea of nature as clockwork and only later came to understand nature as an organism; but even then – even when his conception of nature 'could no longer be reduced to a machine model'[3] – he did not completely abandon such a model. The conjunction 'or', which connects the two terms in the title of her essay, is therefore to be interpreted as inclusive, and this title also finishes on a full stop and not a question mark.

According to Kuhn, Kielmeyer stands in the history of the natural sciences between two epochs (Enlightenment and Idealism) and their different models of nature (mechanism and organism) and cannot be assigned to one or the other epoch or model of nature without loss:

> His personal development is part of the history of natural science and represents the period which marks the step from the Enlightenment with its non-specific mechanistic natural science, determined by physics, to that of German Idealism, in which biology was privileged. But the tension in Kielmeyer's view of nature between 'clockwork' and 'organism' ... finds no solution in itself. The notion of machines had become too narrow for him, he only partially implemented Kant's 'Copernican turn'. Kielmeyer's 1793 speech stands at the crossroads between a static view of nature, in which existence in the best of all worlds is regulated by the meaningful interplay of all forces in space and time, and a more dynamic view, in which everything develops towards greater perfection. Schelling's remark that a new epoch of natural history begins at this point is therefore understandable. Kielmeyer starts from the balance of forces, comes to the idea of development and tries to combine both in his system of organic forces. Now, however, he refuses to be classified as a natural philosopher or even as the father of natural philosophy; empiricism remains his rule; he does not accept principles

that precede experience. His conception of nature as an organism – even if he includes development – remains bound to the equilibrium-conception of a purposeful interplay of forces.[4]

It is to Kuhn's credit that she named the two traditions within which Kielmeyer stands, distinguished them from one another and, in the end, showed that the tension between the metaphor of the clockwork and that of the organism has no solution in Kielmeyer. To repeat Schelling's statement, she ushers in a new epoch of research on Kielmeyer that does not treat him anachronistically as a biologist, as the zoologist Heinrich Balss had done forty years earlier. In his essay 'Kielmeyer as a Biologist',[5] Balss uses terms from his time (e.g. descent theory) to describe Kielmeyer's position as a biologist that are not those of Kielmeyer. There is, to speak with Georges Canguilhem, no complete identity of question and research intention, nor of guiding concepts and systems of concepts, that holds between Kielmeyer's time and biology.[6] In Balss' paper, Kielmeyer's biological views are produced out of a reconstruction based on the understanding of biology around 1930.

But Balss should have known better, since he follows Max Rauther,[7] who in his essay on those who 'excavate "anticipations"' warns against a too modern interpretation of older texts:

> If we look back from the late stages of the development of science to the times when the development of the possible material of experience was still in its infancy and just began to form guiding principles, we are tempted to interpret the terms and opinions of the old writers into expressions which have become flesh and blood for us today in a stricter way – expressions which we tend to consider as more or less definitive. This leads to the search for 'anticipations' of modern theories, which sometimes become rather violent (I need only recall the argument whether Goethe is to be regarded as 'Darwin's predecessor' or not, etc.). For mostly the ideas of the ancients are much more general and arise on quite different presuppositions than those of a later time.[8]

So instead of digging up 'anticipations' of a future biology in Kielmeyer's work, I would like to turn to the presuppositions of his thought. I am not so much interested in the sources Kielmeyer chose for the differentiation of his five organic forces or for the formulation of his system of organic forces[9] as in the questions of contemporary physics from which Kielmeyer develops his thinking and his method of comparison. For it is evident – especially with regard to the 'history of the sayable'[10] – that Kielmeyer was not able to think within the horizon of an already-established biology.[11] He had, instead, to think in the context of the physics of his time. Yet, one has also to consider what Rudolf Stichweh has shown in his groundbreaking study *Zur Entstehung des modernen Systems wissenschaftlicher Disziplinen*: physics as a nineteenth-century discipline has 'relative irrelevance'[12] for eighteenth-century science. Physical knowledge is not oriented towards 'physics', but towards a 'hierarchy of forms of knowledge constituted by the distinction between historical, philosophical and mathematical knowledge'.[13]

Organic Physics

Carl Philipp von Martius long ago pointed out that Kielmeyer saw all the doctrines that he taught as 'parts of physics'.[14] In order to understand this, it is helpful to briefly recall some key points in the development of physics. Using Gunter Lind's work, the origins and lines of development of this physics can be summarized briefly as follows:

> At the beginning of the eighteenth century the terms 'physica' and 'philosophia naturalis' were synonymous. Physics was a part of philosophy, namely that [part] which had to do with the knowledge of the essence and the interrelation of things in nature. The object of physical knowledge was all of nature, animate and inanimate, physical and mental. Nature was thereby understood as God's creation. ... The intention with which nature was viewed was theoretical. Physics wanted to explain nature, both causally and teleologically. It concerned the question of the causes of things and the question of their purpose in the divine world plan. The science of physics, which aimed at explanation, was contrasted with those sciences that were only concerned with the description of nature. *Naturgeschichte*, which was limited to observing, collecting and ordering (later called *Naturkunde*), was often understood as a prerequisite of physics, but lacked its philosophical interest in knowledge.[15]

Lind goes on to explain that, compared to this initial situation, the meaning of the concept of physics and, consequently, the subject matter of physics changed fundamentally over the course of the eighteenth century. Physics broke away from its 'philosophical interest in knowledge' to define itself by means of 'methodical access' to its objects. This changed the 'explanatory concept' it strives for: teleological causes are 'eliminated' and causal causes are 'reinterpreted': 'Cause-analysis is no longer seen as the discovery of the acting potencies, but as the determination of regularities'.[16] Ultimately, this also had far-reaching consequences for the subject matter of physics, which was now 'severely restricted':

> What was beyond methodical access, what could not be grasped experimentally or described mathematically, was excluded; first the doctrine of the soul, then the doctrine of life. On the other hand, the physical parts of applied mathematics were added. Physics became the experiential science of inanimate nature.... Physics then understood itself as the science of the basic laws within the domain of inanimate nature.[17]

It goes without saying – and this is no criticism of Lind's remarks – that this summary is a rational reconstruction of lines of development that necessarily ignores certain asynchronous tendencies. For, while some natural scientists wanted to exclude animate nature from physics, others made great efforts to philosophically and scientifically process the organic parts of a natural history that had until that time only been a matter of history and description, in order to give natural history the status of a natural doctrine or to preserve the realm of the living as a subject of physics. Hermann Schlüter, who has studied the period between 1790 and 1870, defines this organic physics as a science 'which attempts to give biology a form of axiomatization similar to Newtonian physics'.[18]

However, Kielmeyer is less concerned with such axiomatization: He does not bring order to the set of already-existing statements within organic physics, but – by ordering and structuring the set of phenomena in the field of organic nature using five organic forces – he outlines the subject matter for a future organic physics. This was possible, because physics at that time had no clearly defined boundaries. Perhaps it was precisely the vagueness of the boundaries of physics that provided the condition for the possibility of an organic physics, i.e. for that pursuit which disappeared from the corpus of physics at the very moment when physics became established as a discipline. In any case, an indispensable prerequisite for organic physics is that the realm of the organic is to establish itself as independent and different from the inorganic.

This has been achieved since, at least, the tenth edition of Carl von Linné's *Systema naturae* of 1758, which introduced a dichotomy into the three kingdoms of natural history: whereas unanimated natural bodies (the mineral kingdom) are merely composed, animated natural bodies (the plant and animal kingdoms) are organized.[19] Research in the field of conceptual history shows that towards the end of the eighteenth century there was a shift in the semantics of terms[20]: the term 'organism' replaced that of 'organisation' and 'began to establish itself as a technical term for "living beings" in the field of biology'.[21] At this time, an 'unleashing of the concept of "organ" under the idea of life force'[22] occurs and the adjective 'organic' emancipates itself 'from its attachment to the physis-techne analogy; it now serves to identify the living, in contrast to the mechanical, as such'.[23]

The Limits of Physics around 1800: Claims and Reality[24]

Even though the modern sciences were invented between 1572 and 1704,[25] around 1800 it is still not clear what distinguishes and defines physics as a science. Due to the possibility of formulating its results mathematically, mechanics is considered exemplary for all areas of physics.[26] On closer examination, however, it is clear that the form of a mathematical science cannot be realized in all sub-domains of physics. In comparison to mechanics, those physical sub-disciplines that work with qualities as opposed to quantities fell short of such scientific standards (e.g. the study of electricity, whose subject area had yet to be developed).[27] Strictly speaking, around 1800, there was no clarity as to what is meant by physics as a science.

This uncertainty in determining the limits of a physics is confirmed by the article 'Physics' in Johann Samuel Traugott Gehler's famous dictionary of physics. Here, numerous synonyms for physics are given:

> Physics, physical science [*Naturlehre*], natural history, natural science, Physica, Physice, Philosophia naturalis, Physique. This is the name given to the entire

teaching of the nature of the body world, or of the properties, forces and effects of bodies. In the broadest sense of the word, natural science includes everything that has ever been experienced or thought about bodies.[28]

If one subtracts the literal translations of physics into common national languages, the following synonyms remain: physical science (*Naturlehre*), natural history (*Naturgeschichte*), natural science (*Naturwissenschaft*) and *philosophia naturalis*. Each of these names designates 'the entire doctrine of the nature of the corporeal world [*Körperwelt*]', regardless of the differences in their methodological approach to nature.

These differences in approach to the study of nature are also at stake in Gehler's classification of sciences. Our knowledge of natural bodies is either historical or philosophical or mathematical. Natural science is thus divided into three parts: natural history, physics and applied mathematics. All these sciences, each of which deals with the 'nature of the corporeal world [*Körperwelt*]' in a different way, are, first of all, natural sciences in the broader sense. Natural history offers an enumeration, naming and describing substances and natural bodies, including their properties, phenomena and effects. Applied mathematics examines the proportions between the natural bodies. And physics as a natural science in the narrower sense aims at 'developing what the special phenomena have in common, or discovering the laws of nature, explaining the phenomena and events from these laws, and investigating causes and motivating forces [*Triebfedern*]'[29]. What is interesting about this characterization of actual physics is that the first two fields of activity do not require any investigation of causes. In order to discover the laws of nature and explain phenomena by means of these laws, it is not necessary to know the causes of these laws of nature and phenomena. According to Gehler, it is even 'the greatest advantage of modern physics that it is more concerned with determining and correcting the laws than with discovering the causes, i.e. more [concerned] with experience than with speculation.'[30] It is exactly as Lind said: physics is an explanatory science, but it does not necessarily aim at knowing causes. These should also be researched, but how far physics will go in examining them remains uncertain. Thus, physics is open to very different fields of objects, i.e. to organic bodies.

However, physics does still have difficulty in determining these subjects. And the conclusion to these discussions of terms is therefore unsatisfactory when it comes to the definition of the concept of physics: 'That which the teachers of nature are accustomed to presenting under the name of actual physics therefore still has very vague limits.'[31]

According to Gehler, physics does not intend to describe its objects in terms of natural history, but to explain them philosophically. In so doing, it is not only concerned with ideal mathematical structures, but with phenomena from the world of bodies that can be measured in real terms. In its concrete actions, physics begins descriptively with the elaboration of commonalities, then natural laws are established and finally the phenomena are explained by these laws. The philosophical aspect of physics is to explain phenomena or occurrences by natural laws, even if physics has to admit that it cannot explain the phenomena by 'essential' causes in this way. The concept of natural law is also very broadly defined: 'With this name [natural laws] one proves certain general rules drawn from the observations of nature, according to which these or those effects and changes take place in the body-world'.[32]

As a natural science which explains its objects philosophically, physics aims at the 'explanation of phenomena'.[33] As a natural science, it is an empirical science. To explain something physically means 'that one derives the phenomena from the laws of nature'. And, as long as one refrains from speculative considerations, 'from this a quite certain and correct knowledge of nature emerges, which is completely based on experience and induction, as we have had it since Bacon's and Newton's time'.[34]

As a science, physics in the narrower sense is thus a science of phenomena and not of causes. And, as an experiential science, it is nothing other than a science of the experience of nature, or in short: *phenomenology of nature*. Its business is to derive the laws of nature from phenomena and to explain the phenomena from laws of nature – and, according to Gehler, these are only experiential theorems. But knowledge of the laws of nature does not imply knowledge of the effective causes: 'Therefore, knowledge of the laws of nature is not knowledge of the effective causes and the mechanism by which the phenomena are produced in fact. The laws teach only what happens, not how and by what means it happens.'[35] Although the laws of nature are formulated mathematically as far as possible, this is only possible in some areas of physics, such as mechanics or optics. As an inductive science, physics as a whole has various methodological approaches; however, its main tasks include observation, measurement, experimentation and, as far as possible, the mathematical representation of the laws of nature and the formulation of hypotheses. Based on these findings, it does not seem unreasonable to deal with the field of organic matter within the framework of physics, which here presents itself as a phenomenology of nature. Organic physics is physics in exactly this sense and the adjective 'organic' only refers to the specific subject of this physics.

The Organic: Between Physics and Physiology

The methods of physics suitable for determining the organic are also discussed at the time in the context of natural history and physiology.[36] In this vein, the most famous natural history of the eighteenth century, Georges-Louis Leclerc de Buffon's,[37] does not only contain descriptions of natural things, but also gives explanations for the facts, theories and hypotheses there described.[38] Perhaps Albrecht von Haller's preface to the German translation of this natural history[39] was intended, in part, to prepare the reader for such unexpected theoretical components and hypotheses. In any case, it is striking that Haller questions the applicability of mathematical doctrine to natural history and argues for the utility of hypotheses. In so doing, he starts a discussion about what physical methods are most useful for natural history and natural science.

As far as mathematical doctrine is concerned, according to Haller, it is impossible to explain nature mathematically, since most of the moving forces of nature are unknown: 'Can we be expected to be mathematically rigorous, can a sum of concepts be certain when the individual ones are still undetermined?'[40] On the other hand, Haller argues for the 'usefulness of the hypotheses'. These are a heuristic tool for research, through which we can better understand nature. Hypotheses are not yet truths, 'but they do lead to [them]': 'They are the guide that leads to the new and the true, and I can think of no inventor who has not used the same [guide].'[41] The only limitation of hypotheses is that they are only means 'for approaching the truth' and not the truth itself. Hypotheses are therefore not permanent, but always provisional.[42]

In the preface to his early physiology,[43] Haller goes one step further in the direction of 'organic physics'. By referring to specifically organic phenomena that cannot be explained by mechanical forces, he challenges the general application of mechanical explanations. In the case of the animal machine, many phenomena occur that cannot be explained by ordinary mechanical laws – and they all have the same structure: on the side of effects, there are strong movements excited by small causes.[44] According to Haller, the phenomena he enumerates cannot therefore be explained by the moving forces encountered outside the 'animated machines'. Haller does not, however, speak against the mechanical laws encountered outside the animate bodies in general, but he does demand that these laws be applied to organic bodies only if they are consistent with experiments.[45] Haller thus implicitly distinguishes between the moving forces of inanimate and animate nature, although the contrast between the organic and the inorganic itself is not yet explicit at this point: in the concept of the animated machine, these two domains are still harmoniously linked.

With this first systematization of life phenomena that cannot be explained mechanically, Haller goes beyond the explanations he provided in his treatise 'De partibus corporis humani sensibilibus et irritabilibus'.[46] In this speech, Haller had distinguished merely irritable from sensitive parts, which he then separates from the parts that are neither irritable nor sensitive. He does not have a theory as to why certain parts are irritable and others sensitive (and yet others neither irritable nor sensitive), nor does he not wish to promise such a theory, for he is convinced 'that the source of these two forces lies hidden in the innermost structure of the parts, and that it is far too fine to be discovered with the aid of the anatomical knife or the magnifying glass'.[47]

For what follows Haller's qualitative distinction between sensitivity and irritability is important. While phenomena of sensitivity only occur while the body is alive, phenomena of irritability can also be demonstrated on dead bodies. To Georg Ernst Stahl and his followers, Haller can show by experience that irritability is not dependent on the soul. His explanations prove that the phenomenon of irritability does not depend on the organic body being alive: 'the irritability remains when the soul has either left its seat or its communion with the body has been interrupted, and the irritability does not depend on the soul'.[48] Haller asserts this without leaving any room for misunderstanding: irritability is a property of the muscle fibre. And he goes a step further when he says that it is not necessary to specify 'another cause' for it, because, even when it comes to the phenomena of attraction and severity, it is still not necessary to specify another cause.[49]

Haller's definition of irritability as a property of the muscle fibre remains within the limits of a purely phenomenological determination of the observed. The cause of irritability is postulated (it lies in the construction of the fibre and is a property of the muscle fibre), but what causes the observed effects of irritability remains open. Haller explicitly compares irritability with Newton's force of attraction or gravity. In both cases, (previous) experiments are only sufficient to discover and testify to the phenomena of irritability and attraction, but these experiments do not succeed in investigating the physical causes of irritability and attraction. With Haller's comments on irritability as a property of the muscle fibre, a special organic force pertaining to the organic body has been identified for the first time: it is demonstrably organic and yet does not depend on the soul. Further forces will follow. The term 'force' is, as Gehler explains, only a 'general name for everything that strives to produce, change or hinder movement', but the causes of this movement are 'hidden in the deepest darkness'.[50] And, basically, we merely 'help ourselves' with the word force, 'in order to designate all those causes which we must name so often, although their nature remains an inscrutable mystery'.[51] And it is in exactly this sense that the word 'force' is only a 'makeshift word' (*Behelfwort*) for Kielmeyer (p. 32).

Another author who speaks of organic forces and appears in the catalogue of Kielmeyer's library[52] is Johann Nicolaus Tetens. His explanations are of particular interest with regard to the distinction between organic and mechanical forces. For Tetens, organic forces are forces as long as their effects cannot yet be explained by the principles of mechanics.[53]

However, Kielmeyer received the most important impetus for his system of organic bodies from Herder, who liberally uses the concept of organic force in his *Ideas for the Philosophy of the History of Mankind*[54]. It functions here as a generic term for all specifically organic forces. In the second chapter of the third book on the 'Comparison of the Various Organic Forces Acting in Animals', Herder concludes from his explanations (or 'inductions'):

1. In every living creature the circle of organic forces seems whole and perfect; only it is differently modified and distributed in each. In the latter, it is still close to vegetation and is therefore so powerful for reproduction and restoration of itself; in others, these forces decrease as they are distributed into more artificial limbs, finer tools and senses.
2. Above the powerful forces of vegetation, the living muscular stimuli begin to work. They are closely related to those forces of the growing, sprouting, recovering animal fibrous structure; only they appear in an artificially intertwined form, for a more limited, specific purpose of life's action. . . .
3. The more the muscular forces enter the area of the nerves, the more they too are trapped in this organization and overwhelmed for the purpose of sensation.[55]

The results formulated here show that Herder does not consider individual organic forces in isolation from each other. The 'circle of organic forces' is 'differently modified and distributed' in each creature, i.e. the ratios of the organic forces to each other differ in the individual organic bodies: plants, plant-animals, cold-blooded and warm-blooded animals each show different ratios of forces. Herder is thus already gesturing towards a law of compensation of forces, without yet formulating this assumption in terms of regularity. And Herder is also here already considering whether the fibre, muscle and nerve force could not be 'basically one and the same force, which is revealed differently in the fibre, differently in the muscle, differently in the nervous system'.[56] Kielmeyer will come to make similar comments. However, the following passage is determinative on Kielmeyer's project:

> Who would not be delighted should a philosophical anatomist [philosophischer Zergliederer] undertake a comparative physiology of several animals, particularly those close to man, in order to furnish, in accordance with this empirical experience, the distinctive and definitive [festgestellten] forces in relation to the whole organization of the creature? Nature presents us with her work: from the exterior in disguised form [verhüllte Gestalt], a covered-over relation of inner forces.[57]

The importance of this passage can hardly be overestimated, for it contains the 'theme' of Kielmeyer's later 1793 Speech.[58] What is more, Herder's inductions also gave important hints for the content of the law of compensation or for speculation over one fundamental force underlying organic forces. In terms of content, however, as Wolfgang Pross has convincingly shown, there are some significant differences between Herder and Kielmeyer.[59] Yet, if Kielmeyer is independent of Herder in terms of content and terminology when setting up his five basic organic forces, there is nothing to prevent him from following the procedure he describes at the beginning of his speech when determining the different organic forces: he separates the effects perceived in the individual organizations 'into classes according to their similarities and differences' and gave 'these effects or their causes ... the names of different forces'. (see p. 32)

Kielmeyer's System of Organic Forces as a Phenomenology of the Organic

Kielmeyer takes up Herder's theme of a comparative physiology of organic forces,[60] which differs significantly from Haller's work. As the title of the speech indicates, he is concerned exclusively with 'the relations between organic forces in the series of different organizations and ... the laws and consequences of these relations'. And, for this reason, he does not go into the usual physiological details, i.e. the distinction between the various systems of the human or animal body, i.e. the bone, muscle, organ and nervous systems, etc.

Kielmeyer points out that he can only present fragments and asks the audience to postpone judgement on the ideas communicated until the publication of his

'Geschichte und Theorie der Entwiklung der Organisationen' (p. 31).[61] However, this paper was not published during Kielmeyer's lifetime, although Fritz-Heinz Holler did publish a manuscript which is most likely one part of it as, 'Ideen zu einer allgemeinen Geschichte und Theorie der Entwicklungserscheinungen der Organisationen 1793/1794'.

Among the manuscripts published by Holler, there is another important work that can provide information about Kielmeyer's research agenda – the 'plan for a zoology and anatomy of animals or the draft for a comparative zoology' (GS, 22). This draft concerns the determination of possible knowledge about animals. Such knowledge would be either historical (in the sense of enumerating the changes or determining the conditions) or theoretical (in the sense of determining the conditions of changes by ways of laws to the point of determining causes and consequences) (GS, 22). This plan also aims at a broad comparative study of the 'invigorating and driving forces' of the organism, in which forces are again only considered as 'classes of effects with unknown cause'. In sum, Kielmeyer is concerned with a 'dynamics of animals' (GS, 27). The formulation is deliberately chosen, because, in physics, dynamics deals with the 'theory of the forces causing the movement of bodies' (GS, 27). A dynamics of animals would therefore be a theory of the forces that cause the movement of animal bodies – and this project fits nicely into organic physics.

When Kielmeyer conceives the series of organizations as a 'series of successful attempts by nature' to 'present the phenomenon of life and the organism' (GS, 27), it becomes clear that he does not only regard individual organisms as phenomena, but life itself as a phenomenon. He understands this phenomenon of life as a 'sum of fractions', 'which only by summing up give a whole, the integral of life' (GS, 27). It is striking that Kielmeyer at this point falls back on a mathematical vocabulary. He speaks of series, fractions, sums and the integral of life. Above all, the term 'integral' (an integral is something that constitutes a whole and exists in its own right) underlines that Kielmeyer takes life as a whole and as something that exists for itself, and, in this respect, differentiates it from the inorganic. Life therefore follows its own laws and is characterized by its own specifically organic forces.

Kielmeyer therefore focuses on life as a whole and not only on individual living beings. If Foucault is right with his assertion that it is wrong to write stories of biology in or for the eighteenth century, because at that time life itself did not yet exist, but only living beings,[62] then this no longer applies to Kielmeyer. The physics of organic bodies that he projected – in this manuscript he speaks of the 'physics of the animal kingdom' (GS, 28), since it is a matter of comparative zoology – would therefore actually constitute an important stage on the path to life science or biology.

In his classification of organic forces (sensibility, irritability, reproduction, secretion, propulsion), Kielmeyer uses the concept of ability as a synonym for force (GS, 28). Etymologically, the preposition 'ability' ('*vermöge*') also describes nothing but 'the possibility, quality or ability lying in someone, something, which is the reason that something happens or exists'.[63] Since the word 'ability', understood as the power to do something, is a translation of the Latin 'facultas', Kielmeyer's organic forces or abilities could replace the old Aristotelian 'facultates animae'.[64] In any case, animals are classified in terms of these abilities or organic powers. This takes place on the level of classes (here different genera – today one would speak of species – are interrelated), on the level of genera (here different individuals within a genus are interrelated), on the level of individuals and on the level of organs. The comparative study is meant to show what forces are present (existence), in what different ways they manifest themselves (diversity of expressions), how long they last (duration), how strong they are (intensity) and how they relate to each other (relationship to each other). Subsequently, the laws, causes and consequences are to be determined. The established classes of effects or forces are then compared with each other again to discern whether they can be traced back to one or two unprovable preconditions or causes. All this corresponds in detail to the line of argumentation given in the 1793 Speech.[65]

Here too, the integral of life – or the life of the great machine of the organic world (p. 47) – is approached by way of phenomenology. The phenomena of nature are interpreted as a system of interactions linked in space and time (p. 47). These interactions have both a synchronous and a diachronic dimension: in animate nature, they are found at the levels of organs, individuals and genera, and Kielmeyer sometimes focuses on their synchronous dimension of spatial interaction and sometimes on the diachronic dimension of temporal development. In the end, all these interaction-systems converge into one final great system, the life of the great machine of the organic world. In this system, the effects of the individuals from one species interact with the opposite effects of the individuals from all other species. In this respect, Kielmeyer does not speak of an interaction of effects, but of the fact that the effects are chained together (*zusammenverkettet*) (p. 47). This machine progresses on a path of development which Kielmeyer illustrates with the mathematical image of a parabola that never circles in on itself (p. 30). What impresses Kielmeyer most, is the fact that this organic nature 'as a whole always remains the same, and nature's quiet course goes on unhindered', which is why 'the question of the causes and forces by which this effect is obtained' is also an obvious one.

Kielmeyer asks himself three questions: 1. which forces operate in animate nature, 2. which power relations manifest in individuals and how do these relations change in the series of the different organizations and 3. how can the course and existence of the organic world be explained from laws concerning changes of relations. According to Kielmeyer, the focus of the investigation is on answering the second question (p. 31).

To answer the first question, Kielmeyer separates the effects perceptible in the individual organisms 'into classes according to their similarity and differences' and proves 'these effects or their causes, so long as they aren't better known, with the makeshift word [*Behelfwort*] "forces"' (p. 32). In this way, he comes to distinguish five organic forces: sensibility, irritability, reproductive power, secretion and propulsive power.

Kielmeyer introduces his answer to the second question with a preliminary consideration. He asks himself what standard by which 'they can all be measured and quantitatively compared' (p. 33). Since there is no general standard for the intensity of these forces, he suggests that one might examine the observed effects by way of 'number or frequency' as well as 'multiplicity', 'amount of the resistance' and 'permanence of effects' (p. 33). With regard to these parameters, he then succeeds in determining the different relations of organic forces in the various organizations. In so doing, 'at first each particular force will be kept isolated from similar ones in other species' (p. 33) and then he considers them 'in combination with other forces' (p. 33).

Hence, Kielmeyer attains individual laws for sensibility, irritability and reproductive power, from which he then derives the more general law of compensation, which states that, in the series of organizations, sensibility is displaced by irritability and reproductive power and irritability is finally displaced by reproductive power: 'the more one of these forces is cultivated on one side, the more it is neglected on the other' (p. 41).

The answer to the question of 'the course and continued existence' of the organic world has two parts. First, Kielmeyer envisages the existence of the organic world as guaranteed by the fact that opposing forces compensate each other. There is a balance between the destructive and the maintaining forces. For example, there is an equilibrium between insensitive and non-irritable plants and their sensitive and irritable predators insofar as the reproductive power of the plants is so strong that the number of offspring is large enough to avoid the predators' behaviour leading to the extermination of the plants. Secondly, the course of nature is also a product of this balance between destructive and sustaining forces. Kielmeyer admits, however, that, in the case of an 'over-preponderance in the destructive forces' (p. 44), it is quite possible that genera could become extinct.

Kielmeyer concludes his remarks with the hypothesis that these relations of forces could possibly be derived from only one single force, 'in the same way that light appears split into different rays and these rays are mixed into infinitely different proportions'. By this force 'the smallest organ, right up to the most complex and immense machine, is set in motion' (p. 45). Kielmeyer speculates that this force was perhaps originally awakened by light, as it is still maintained by it.

Classification of phenomena as effects of unknown forces, derivation of laws of nature from the phenomena of nature, generalization of these laws and a sparing use of hypotheses and explanations of causes – Kielmeyer makes use of the methodological framework of organic physics. First, the phenomena or occurrences of nature are examined and assigned to individual organic forces according to classes of effects, then laws are derived from the phenomena – first, one for each of the three principal forces, then a further generalization is introduced, and the compensation law reduces the individual laws to a common denominator – and finally Kielmeyer even puts forward a hypothesis about the fundamental force – first, in the sense that all organic forces can be derived from it, and then also in the sense that this fundamental force itself is in turn derived from a mechanical force.

Kielmeyer's system of organic forces and his comparative phenomenology of the organic can thus be fully explained in the context of organic physics. The term 'organic physics', which fits the context of physics around 1800, therefore seems to me suitable to classify and describe Kielmeyer's achievements. This organic physics does not yet aim at axiomatization, because Kielmeyer does not intend to bring order to the set of statements for an already-established discipline of 'organic physics'. That would be too much to ask. However, by ordering and structuring the set of phenomena in the field of organic nature and by looking at life as a whole, he outlines the field of organic physics – a field which biology as a science will eventually come to inhabit.

Notes

1 The present text goes back to an essay first published in the anthology, *Physik um 1800* edited by Olaf Breidbach and Roswitha Burwick: T. Bach, 'Organische Physik als vergleichende Phänomenologie des Organischen. Anmerkungen zum wissenschaftshistorischen Ort von Carl Friedrich Kielmeyers System der organischen Kräfte', in O. Breidbach and R. Burwick (eds), *Physik um 1800. Kunst, Naturwissenschaft oder Philosophie?* (Munich: Fink, 2012), 187–222. This version reformulates the earlier thesis that Kielmeyer 'does not think within the horizon of an already-established biology, but in the context of an organic physics or physics of organic bodies' and therefore 'his system of organic forces must be placed in the context of physics around 1800' (ibid., 191).
2 D. Kuhn, 'Uhrwerk oder Organismus. Karl Friedrich Kielmeyers System der organischen Kräfte', in *Nova Acta Leopoldina*, Neue Folge 36 (1970), 157–67.
3 Ibid., 163.
4 Ibid., 167. See also D. Kuhn, 'Der naturwissenschaftliche Unterricht an der Hohen Karlsschule', in *Medizinhistorisches Journal* 11 (1976), 319–34.
5 H. Balss, 'Kielmeyer als Biologe', in *Sudhoffs Archiv* 23 (1930), 268–88.
6 G. Canguilhem, *Wissenschaftsgeschichte und Epistemologie. Gesammelte Aufsätze*. (Frankfurt am Main: Suhrkamp, 1979), 35.
7 M. Rauther, 'Ungenutzte Quellen zur Kenntnis K. F. Kielmeyer's', in *Besondere Beilage des Staats-Anzeigers für Württemberg* 6 (14 May 1921), 113–22.
8 Ibid., 115.
9 Cf. T. Bach, *Biologie und Philosophie bei C. F. Kielmeyer und F. W. J. Schelling* (Stuttgart-Bad Cannstatt: Frommann-Holzboog, 2001), 89–198. On the influences of Johann Gottfried Herder and Eberhard August Wilhelm Zimmermann, see W. Pross, 'Herders Konzept der organischen Kräfte und die Wirkung der "Ideen zur Philosophie der Geschichte der Menschheit" auf Carl Friedrich Kielmeyer', in K. T. Kanz (ed.), *Philosophie des Organischen in der Goethezeit: Studien zu Werk und Wirkung des Naturforschers Carl Friedrich Kielmeyer (1765-1844)* (Stuttgart: Franz Steiner, 1994), 81–99.
10 A. Landwehr, *Geschichte des Sagbaren. Einführung in die Historische Diskursanalyse* (Tübingen: Diskord, 2001).
11 This does not exclude the possibility that Kielmeyer's 1793 System can, with the necessary limitations, be interpreted as an 'early systems-programme of biology', see Bach, *Biologie und Philosophie*, 87 and J. H. Zammito, *The Gestation of German Biology: Philosophy and Physiology from Stahl to Schelling* (Chicago: University of Chicago Press, 2017), Chapter 9.

12 R. Stichweh, *Zur Entstehung des modernen Systems wissenschaftlicher Disziplinen: Physik in Deutschland 1740-1890* (Frankfurt am Main: Suhrkamp, 1984) 14.
13 Ibid., 15.
14 C. F. P. von Martius, 'Denkrede auf Carl Friedrich von Kielmeyer', in Martius, *Akademische Denkreden* (Leipzig, 1866), 11.
15 G. Lind, *Physik im Lehrbuch 1700-1850* (Berlin: Springer, 1992), 2.
16 Ibid., 3.
17 Ibid.
18 H. Schlüter, *Die Wissenschaften vom Leben zwischen Physik und Metaphysik. Auf der Suche nach dem Newton der Biologie im 19. Jahrhundert* (Weinheim: VCH, 1985), 215.
19 G. Toepfer, '"Organisation" und "Organismus" – von der Gliederung zur Lebendigkeit – und zurück? Die Karriere einer Wortfamilie seit dem 17. Jahrhundert', in M. Eggers and M. Rothe (eds), *Wissenschaftsgeschichte als Begriffsgeschichte. Terminologische Umbrüche im Entstehungsprozess der modernen Wissenschaften* (Bielefeld: de Gruyter, 2009), 83–106.
20 See the articles 'Organ', 'Organisation' und 'Organismus' in G. Toepfer, *Historisches Wörterbuch der Biologie. Geschichte und Theorie der biologischen Grundbegriffe*, 3 vols. (Stuttgart: Springer, 2001/2011), 2: 746–53, 754–76 and 776–842.
21 Toepfer, '"Organisation" und "Organismus"', 102.
22 J. H. Wolf, *Der Begriff 'Organ' in der Medizin. Grundzüge der Geschichte seiner Entwicklung* (Munich: Fitsch, 1971), 38–9.
23 Theodor Ballauf and Eckart Scheerer, 'Organ', in J. Ritter et al. (eds), *Historisches Wörterbuch der Philosophie*, vol. 6 (1984), 1320.
24 The main source for my reconstruction of physics around 1800 is Gehler's *Dictionary of Physics*. The first four volumes were published between 1787 and 1791. Cf. J. S. T. Gehler, *Physikalisches Wörterbuch oder Versuch einer Erklärung der vornehmsten Begriffe und Kunstwörter der Naturlehre mit kurzen Nachrichten von der Geschichte der Erfindungen und Beschreibungen der Werkzeuge begleitet in alphabetischer Ordnung*, 6 parts (Leipzig, 1787–96). – Kielmeyer possessed this dictionary – see F. F. Autenrieth, *Verzeichniß der von Herrn Dr. v. Kielmeyer . . . hinterlassenen Bibliothek* (Stuttgart, 1845), 24: '556. Gehler, physikalisches Wörterbuch. 6 Bde. m. Kpfrn. Leipz. [1]787-96.'
25 Cf. D. Wootton, *The Invention of Science. A New History of the Scientific Revolution* (London: Penguin, 2015).
26 A sign of this is that the mechanistic model of thought had also established itself in the field of medicine and in the explanation of the phenomena of the living. Cf. I. Jahn, *Grundzüge der Biologiegeschichte* (Jena: Fischer, 1990), 159–218.
27 Cf. Gehler, *Physikalisches Wörterbuch, Erster Theil von A bis Epo*, 719–831.
28 Gehler, *Physikalisches Wörterbuch, Dritter Theil von Liq bis Sed*, 488.
29 Ibid., 489.
30 Ibid., 325,
31 Ibid., 490.
32 Ibid., 325.
33 Ibid., 495.
34 Ibid., 457.
35 Ibid., 324.
36 On natural history see W. Lepenies, *Das Ende der Naturgeschichte. Wandel kultureller Selbstverständlichkeiten in den Wissenschaften des 18. und 19. Jahrhunderts* (Frankfurt am Main: Suhrkamp, 1978); on physiology, see J. Jantzen, 'Physiologische Theorien', in

M. Durner et al., *Wissenschaftshistorischer Bericht zu Schellings naturphilosophischen Schriften 1797-1800* (Stuttgart: Frommann-Holzboog, 1994), 373–668.
37 G. L. L. de Buffon, *Histoire naturelle générale et particulière* (Paris 1749–89).
38 See H.-J. Rheinberger, 'Buffon: Zeit, Veränderung und Geschichte', in *History and Philosophy of the Life Sciences* 12 (1990), 203–23; Bach, *Biologie und Philosophie*, 227–32.
39 On Haller, see R. Toellner, *Albrecht von Haller. Über die Einheit im Denken des letzten Universalgelehrten* (Wiesbaden: Steiner, 1971); H. Steinke et al. (eds), *Albrecht von Haller. Leben – Werk – Epoche*, 2nd edn (Göttingen: Wallstein, 2009).
40 A. von Haller, 'Vorrede', in G.-L. Buffon, *Allgemeine Historie der Natur nach allen ihren besonderen Theilen abgehandelt . . ., Erster Theil* (Hamburg und Leipzig, 1750), xiii.
41 Ibid., xiv; see S. A. Roe, 'Anatomia animata. The Newtonian physiology of Albrecht von Haller', in E. Mendelsohn (ed.), *Transformation and Tradition in the Sciences. Essays in Honour of I. Bernard Cohen* (Cambridge: Cambridge University Press, 1984), 275–6.
42 Haller, 'Vorrede', XVI. – On hypotheses see the article 'Hypothesen' in Gehler, *Physikalisches Wörterbuch, Zweiter Theil von Erd bis Lin*, 677: 'One can therefore by no means doubt the great usefulness and indispensability of hypotheses in physics. Where one has no other means of explaining nature, they are the only link by which one can link several occurrences and be guided on the way to an expedient multiplication of observations and experiments, even to the discovery of the true cause. . . . Good hypotheses, even if they are not the truth itself, make the connection between the events more sensible, provoke experiments and discoveries which one would not have thought of without them, and incessantly encourage the unparthean observer to new tests which almost always teach something useful.'
43 A. von Haller, *Anfangsgründe der Phisiologie des menschlichen Körpers* (Berlin, 1759–76); Kielmeyer owned the Latin edition, see Autenrieth, *Verzeichniß der Bibliothek*, 6: '116. – elementa physiologiae corporis humani. VIII Tomi et auctarium ad Halleri elementa physiologiae, Fasciculus I.-III. et IV. pars. 1. Laus. 757-80.'
44 Haller, *Anfangsgründe*, 16 (unpaginated).
45 Ibid.
46 A. von Haller, 'De partibus corporis humani sensibilibus et irritabilibus', in *Commentarii Societatis Regiae Scientiarum Gottingensis*, vol. 2 (Göttingen, 1753). See A. von Haller, 'Von den empfindlichen und reizbaren Theilen des Menschlichen Körpers. Eine Vorlesung: Am 22ten April 1752, in der Kön. Gesells. der Wiss. zu Göttingen', in *Sammlung kleiner Hallerischer Schriften* (Bern, 1772).
47 Haller, 'De partibus', 5.
48 Ibid., 91.
49 Haller, 'Von dem empfindlichen und reizbaren Teilen', 92.
50 See Gehler, *Physikalisches Wörterbuch, Zweyter Theil von Erd bis Lin*, 796.
51 Ibid. 796 f.
52 Autenrieth, *Verzeichniß der Bibliothek*, 1143: 'Tetens, philosophische Versuche über die menschliche Natur und ihre Entwickelung. 2 Bde. Leipz. 777.'
53 J. N. Tetens, *Philosophische Versuche über die menschliche Natur und ihre Entwickelung*, 2 vols. (Leipzig, 1777), 2:684. See Bach, *Biologie und Philosophie*, 145–8.
54 J. G. Herder, *Ideen zur Philosophie der Geschichte der Menschheit*, 4 vols. (Riga und Leipzig, 1784–91). The work is not listed in Autenrieth, *Verzeichniß der Bibliothek*. On Herder and Kielmeyer, see Zammito, *Gestation*, Chapter 9. See also I. Schumacher, 'Karl Friedrich Kielmeyer, ein Wegbereiter neuer Ideen. Der Einfluß seiner Methode des Vergleichens auf die Biologie seiner Zeit', in *Medizinhistorisches Journal* 14 (1979), 99;

K. T. Kanz, 'Einführung', in Kielmeyer, *Über die Verhältnisse der organischen Kräfte untereinander, in der Reihe der verschiedenen Organisationen, die Gesetze und Folgen diese Verhältnisse*, ed. K. T. Kanz (Marburg: Basiliken-Presse, 1993), 35–41; Pross, 'Herders Konzept'; Bach, *Biologie und Philosophie*,150–62; and S. Schmitt, *Les forces vitales et leur distrubution dans la nature* (London: Brepols 2006), 19–25.
55 Herder, *Ideen*, Erster Theil, 124–5.
56 Ibid., 112.
57 Ibid., 125–6, as translated by Zammito in *Gestation*, 250.
58 See Kanz, 'Einführung'; Pross, 'Herders Konzept'; and Zammito, *Gestation*.
59 Pross, 'Herders Konzept'.
60 According to Larson, Kielmeyer's essay 'is best read as a physiological version of the economy of nature'(J. L. Larson, 'Vital forces: Regulative Principles or Constitutive Agents? A Strategy in German Physiology, 1786-1802', in *Isis* 70 (1979), 242).
61 The title, with its connection of history and theory, recalls Kant's *Allgemeine Naturgeschichte und Theorie des Himmels* of which Kielmeyer had two editions in his library. See Autenrieth, *Verzeichniß der Bibliothek*, 31: '707. – allgemeine Naturgeschichte und Theorie des Himmels. Zeitz. 798. 708. – dasselbe. Zeitz. 808.'
62 M. Foucault, *Die Ordnung der Dinge. Eine Archäologie der Humanwissenschaften* (Frankfurt am Main: Suhrkamp, 1974), 168.
63 G. Drosdowski (ed.), *Duden. Das große Wörterbuch der deutschen Sprache in sechs Bänden*, Vol. 6, Sp-Z (Mannheim: Duden, 1981), 2763.
64 Bach, *Biologie und Philosophie*, 198.
65 See Bach, *Biologie und Philosophie*, 43–62, Bach, *Organische Physik*, 215–9.

Chapter 9

THE PATH OF THE GREAT MACHINE: KIELMEYER'S ECONOMY OF EXTINCTION

Lydia Azadpour

Kielmeyer announces the main aim of his 1793 Speech as follows:

> The greatest effect, the one which activates our attention most, must be the fact that irrespective of the many conflicting forces in it, so far as we can tell, as a whole nature always remains the same, and nature's quiet course goes on unhindered; the question of the causes and the forces by which this effect is obtained is the most pressing to us.
>
> p. 31

With this claim, Kielmeyer assigns a substantial explanatory role to the distribution of organic forces. As the quotation indicates, there are two aspects he takes this to include:

1. The distribution of proportions of organic forces explains the ongoing *persistence* of the organic world, and
2. The distribution of proportions of organic forces explains the history of the organic world, as a *course* involving change over time.

Accordingly, the conceptual apparatus Kielmeyer sets up to understand the distribution of forces will be critical for drawing out the implications of his account of nature's 'course' or history. For Kielmeyer, this apparatus involves articulating a 'natural economy'[1] using the notion of equilibrium. The guiding question of this chapter is therefore: What does this equilibrium model of the natural economy imply about organic nature's history? By focusing on the issue of extinction events, my analysis will illuminate some of the implicit presuppositions and possible implications of Kielmeyer's developmental history of organic nature. Whilst scholars have recognized that for Kielmeyer the cause of the course and persistence of the organic world is the distribution of various organic forces, the precise formulation of this cause in Kielmeyer's speech has received less attention than the thematic results for which it is more often discussed. Much of the debate has focused on whether or not he used teleological reasoning,[2] whether or not he introduced a version of the phylogenetic law,[3] his role in the formation of biology as its own discipline,[4] and whether he was an evolutionist.[5] Because of these differing focuses and goals, most authors who address Kielmeyer summarize his results but do not go into the details of his account of the distribution of force, though this formulation bears on these issues.

The first section (I) on the economy of organic forces will explain the equilibrium model Kielmeyer uses to make sense of the distribution of forces, in order to address the first point above, the persistence of the organic world.

In the second section (II), I argue that, in 1793, Kielmeyer not only admitted that extinction events were going to happen, if they had not already, but also gave an account of these events internal to the sphere of organic forces.

I will suggest that Kielmeyer's emphasis on balance in the proportion of forces, rather than in populations or in the numbers of individuals, as was common in discussions of natural economy before him, affords him a greater flexibility in his account. Because the distribution of forces is able to shift across species in time, Kielmeyer can explain significant historical changes, such as species extinctions, which he puts down to the initial distribution of forces between kinds, whilst at the same time being able to claim that nature 'remains the same' inasmuch as the total quantity of organic force remains constant.

In the final section (III), the results of the investigation will be used to consider Kielmeyer's wider commitments regarding nature's history. I am particularly interested in the potential implications for Kielmeyer's historicization of nature. Bersier and others have claimed that Kielmeyer is responsible for a "temporalisation" or "animation" of the great chain of being.[6] This mode of argumentation incorporates Kielmeyer into the scheme first discussed in detail by Lovejoy,[7] which holds that a transition from a hierarchical to a temporal chain of being is a key trajectory in the history of thinking about the natural order. Whilst I agree with these scholars that Kielmeyer introduces an important historical path for organic nature in his thinking, I want to note that a model of understanding the organic world on the basis of the distributions of forces between different species does not of itself guarantee a dynamic worldview. Secondly, I suggest that the role of some kind of 'chain of being' in Kielmeyer's writings is open to different possible interpretations.

(I) The Persistence of Organic Nature

This section sets out the components of Kielmeyer's view of the 'cause' of the persistence (and course) of nature. His explanation makes use of the notion of an equilibrium between preserving and destroying forces, sustained by the rules followed by the distribution of organic force. I will first introduce the notion of equilibrium, then its underlying components: organic forces, the laws governing their distribution, and the idea of compensation or offsetting.

According to the view Kielmeyer presents in the 1793 Speech, although many conflicts can be observed in organic nature's forces, these conflicts do not seem to lead to organic nature's diminution or wholescale annihilation. For Kielmeyer, this is something that requires an explanation. Kielmeyer's model of the economy of nature[8] makes use of the concept of 'equilibrium' to describe a balance between 'destructive' and 'preserving' forces:

> Because when comparing the distribution of forces in all organisations, no very marked dissimilarity was produced, and the more excellent force of one was almost always offset by such an excellent force of the other, which, in the case that the first worked towards destruction, equally necessarily entailed preservation, then the continued existence of the organic world must have emerged from this equilibrium of the destructive forces and the forces of preservation, and so too its course, in so far as the mutually balancing forces are always different and express themselves in different ways.
>
> p. 43

This balance between destruction and preservation seems in turn to be made up of a prior, more foundational distribution of organic forces, of which it is an effect.

The notion of balance among living beings had of course been used before Kielmeyer, but as Toepfer notes, the terminology he used to discuss the persistence of organic nature – *Gleichgewicht* – had been introduced to German in 1730 to translate William Derham's use in his *Physico-Theology* (1713).[9] I mention this connection here because Kielmeyer follows Derham's intention of using this concept to explain the *maintenance* of the organic world. Derham wrote that:

> The Balance of the Animal World is, throughout all Ages, kept even, and by a curious Harmony and just Proportion between the increase of all Animals, and the length of their Lives, the World is through all Ages well, but not over-stored.[10]

Derham proposed the idea of an equilibrium among living creatures in a theological context. In his account, various natural phenomena (e.g. differing rates of reproduction among different species with different longevities standing in different predation relations to each other) are understood to be part of God's plan to preserve this balance.[11] Jozef Keulartz notes of Derham's ideology that 'in physico-theology the "struggle for life" is portrayed as an integral and essential element of natural harmony'[12], that is, the conflict created by every animal fighting for survival is productive of a (relatively static) 'harmony' because of the population dynamics of species – governed by their species' longevity, predation rates and reproductive rates. That no kind is considered 'over-stored' in Derham also reveals that he takes there to be a 'correct' number of individuals for each kind; in his compensation mechanism nature tends to maintain a certain order of things in which every species is kept relatively constant in its numbers of individuals.

In Kielmeyer, the concept has the same purpose – explaining the maintenance of the organic world – but the way he sets up his version of this conceptual apparatus has quite different results. The effect that Kielmeyer seeks to explain with his conception of equilibrium is not of the numbers of individuals in populations, as in Derham's account, but rather the quantity of organic force in distribution at a given moment. The concept of equilibrium affords Kielmeyer the possibility for understanding the constancy of a system – the organic world – throughout or despite internal changes to that system. As Toepfer describes it, equilibrium has two key aspects: 'in the attribution of equilibrium the system is awarded a certain latitude of change, without the system itself becoming a different one'.[13] Kielmeyer's equilibrium of preserving and destroying forces is made possible by the various combinations of proportions of organic force that can be tolerated in his system.

(a) Organic Forces

So far, this explanation is quite abstract: it is worth looking more closely, then, at how this apparent equilibrium of preserving and destroying forces is sustained. We can say provisionally that the distribution of certain combinations and amounts of different expressions of organic force to different species of organisms forms an equilibrium of preserving and destroying forces. But what are these 'organic forces' under consideration?

Kielmeyer uses the term 'force' in three ways. According to his initial definition, each of the following terms – sensibility, irritability, reproduction, secretion, and propulsion[14] – describes a category of *effects* that we perceive in organisations: sensibility, the reception of sense perceptions; irritability, response to nervous stimulation; reproduction, the regenerative and reproductive abilities of organisations; secretion, the ability of plants and animals to secrete fluids in particular places of their organisation; and propulsion, the ability of the organisation to move. Second, he also says that these forces can denote the *underlying, unperceived cause* of these effects. When describing each force, he writes that each is a certain 'capacity': e.g. irritability is defined as 'the capacity of some organs – above all the muscles – to contract on occasion of excitement and produce movements' (p. 32). Thirdly, Kielmeyer not only uses the term 'force' to refer to powers and their actualizations, but also sometimes to *the organ that possesses the power*,[15] as for example, in his discussion of reproductive organs. In any case, it is clear that what is essential in something being judged an expression or bearer of organic force is not down to the kind of matter at work but rather its involvement in carrying out a function.

So, Kielmeyer uses the term 'force' quite loosely. Because of the breadth of his use of the term, the ontological status of the forces Kielmeyer identifies is unclear – it seems deliberately so. This is underscored by his claim that his use of the concept of force is 'makeshift' (p. 32), and that he proposes five forces unless they are somehow able to be 'cancelled by a higher understanding' (p. 32), suggesting that their categorization into five kinds is also not absolute, echoing his more general provisional attitude to his theorizations. He remains open to the possibility that we may perceive differences where there are in fact expressions of a unified power of some kind; later in the speech he speculates that this apparent multiplicity is at root one force that differentiates itself and then forms different mixtures, proportions of different expressions.[16]

It is worth adding that Kielmeyer investigates proportions of organic force at several levels in his account: at the smallest scale, within an individual member of a species, and at the broadest of the scales discussed, the distribution of forces between different species kinds. It is this level, of comparison between species that carries the most explanatory weight regarding the persistence of the organic world in his account. Laws regulate the various possible proportions of different kinds of organic force, such that these proportions do not overly disrupt the balance between powers of preservation and destruction. The main preoccupation of the 1793 Speech is working out some of these laws that constrain the relative proportions of forces among different species of organisms.

(b) Laws Regulating the Distribution of Proportions of Organic Force

Two questions inform the laws that Kielmeyer proposes regarding the distribution of forces: what the proportions of forces in different species are, and how these proportions vary along the 'series' of organisations.

In answering these questions, Kielmeyer introduces some presuppositions. First, Kielmeyer's comparison of proportion along a series of organisations requires that there can be some measure by which expressions of the organic forces outlined above can be compared in different individuals and different species. It also requires there is a series of organisations at all, and that these individuals can be grouped into distinct kinds.

The quantifiable aspects of these phenomena that make such a comparison possible, according to Kielmeyer, are the number of effects, the frequency of effects, the permanence of effects, and the amount of resistance that they are opposed to (pp. 35–6). But many of the factors he begins to list as evidence of greater or lesser expenditures are less clearly placeable, for example, the uniformity of build of an animal form requires less force expenditure than a diverse build.[17] In any case, in Kielmeyer's account, observations of these phenomena – though not precisely mathematized – can be quantitatively compared both in individual organisms (at different times) and between species.

What kind of series of organisations does Kielmeyer assume? It is certainly hierarchical insofar as entities are classed as above or below each other: Kielmeyer continually refers to 'lower' and 'higher' or 'more perfect' species in a series ('*Reihe*', '*Stufenfolge*') of organisations. The most concise statement of the variation of proportions along Kielmeyer's series is the following:

> in the series of organisations, sensitivity is gradually displaced by excitability and the reproductive force, and in the end, irritability too gives way to the latter: the more one is increased, the less there is the other, and sensibility and reproductive force tolerate one another the least. Further, the more one of these forces is cultivated on one side, the more it is neglected on the other.
>
> p. 41

There are two hierarchies at work here: the series of species kinds, and the hierarchy of value of different expressions of organic force.

Regarding forces, Kielmeyer takes sensibility to have the greatest value, irritability the next, and reproduction the least. This is reflected in scarcity of distribution in the organic world of the more valuable expressions.

Regarding the series of species, it seems organisms with the greatest proportion of sensibility would be the highest, with irritability, and finally, reproduction, taking up a higher proportion in the 'lower' and still 'lower' organisms. This is how Schelling interprets Kielmeyer – in *On the World Soul*, Schelling praises Kielmeyer for forming a 'graded series'[18] of living beings according to the functions they possess, and in the *First Outline of a System of the Philosophy of Nature* (though Schelling credits Blumenbach and Herder with Kielmeyer's main ideas) he favourably describes Kielmeyer's project as forming a comparative physiological series where 'in the series of organisms, sensibility is displaced by irritability, and as Blumenbach and Sömmering have proven, by the force of reproduction'[19]. Presumably Schelling has in mind relative proportions here – with there being a greater proportion of sensibility in 'higher' organisms, and a higher proportion of reproductive force in lower ones. And it is true that Kielmeyer expresses this view, as we can see from the quote above. But to stress the question of proportion of sensibility alone is not to acknowledge the emphasis Kielmeyer places on the role of diversity of expression in his system of valuation. In Kielmeyer's account, what is important in the implied series of beings is not merely quantity of sensibility, but the diversity of expressions of force that greater quantity of sensibility seems to facilitate. The valuation driving this order seems to result from a commitment to the classical idea of harmony – though not explicitly formulated as a principle – inherited perhaps from Leibniz and the Wolffian school.[20] Insofar as there is a hierarchy of organisms, the most perfect are deemed those with the most differentiated capacities, both within the variety of manifestations between forces, and also forms of a single force. For example, of sensation, Kielmeyer writes that 'in the series of these formations, the ability to obtain manifold, distinguishable classes of sensations is gradually more restricted from humans downwards' (p. 33).

Whatever can contain the most oppositions unified within itself, or the most distinct yet interdependent differences, is the 'highest' animal in Kielmeyer's system of valuation. Again when discussing irritability, Kielmeyer claims that lower organisms can express irritability more independently in each of their parts (for example, when one is cut off), whereas higher expressions can't because their systems are more interdependent. We might call this the 'most organised' insofar as it is able to unify the most differentiated capacities into one individual (or kind of individual). To assign the statuses of 'higher' or 'lower' seems to be a valuation based on a judgement about the more 'organised' beings – where organised refers to the complexity of differentiations unified in an 'organic machine'. The emphasis Kielmeyer places on the complex harmony that the equilibrium forms suggests a connection with Kielmeyer's more general framing in the speech, echoing the opening of the address, where he emphasizes the wonder warranted by the 'multiplicity, manifoldness, and harmony' of effects in organic nature, and the role of this aesthetic reaction in instigating the study of these phenomena.

The account of hierarchy I have just given helps explain the relative distribution of forces among kinds in greater detail. At the most general level, Kielmeyer proposes an offsetting mechanism for conceptualizing the relative distribution of forces among species kinds, in which 'no very marked dissimilarity was produced, and the more excellent force of one was almost always offset by such an excellent force of the other' (p. 43).[21] The offsetting or compensation economy that Kielmeyer describes thus makes equivalences of cost between deployments of different kinds of force. In order to establish these equivalences, Kielmeyer must presuppose the above-mentioned hierarchy of value of the organic forces, such that he can say that a higher quantity of reproductive force can be offset by sensibility. Putting aside for the moment the complications arising from the fact that Kielmeyer only deals at length with three of the five forces he identifies – he does not really explain how exactly propulsion and secretion fit into the laws he posits regarding the distribution of forces – it seems clear that in a species the value of sensibility is greatest, followed by irritability, and reproduction requires the greatest deployment to make up for paucity of the other two. This valuation seems to be drawn, then, from observation of the success of various species, and their resources for maintaining their existence. The diversity of possible expression is vast, and Kielmeyer's laws for possible combinations of proportions of forces allows for quite some latitude in how the forces are manifest.

We can see that through the conceptual apparatus of this 'offsetting' economy, Kielmeyer is able to think about the maintenance of the organic world, as Derham did, but as sustained by an equilibrium of force (via the distribution of organic powers in species). What Kielmeyer seeks to explain is not the numbers of individuals in populations, as Derham's account did, but rather, his units of consideration are the expressions of quantities of organic force in distribution at a given moment. This has the consequence that he is able to entertain more significant changes within his system of organic nature: changes at the level of force in Kielmeyer can affect not only the numbers of individuals in a species, but also the number of species themselves, without a disruption of the equilibrium of force. As we will see in what follows, this formulation opens up the possibility that the apparent equilibrium could be maintained by very different distributions of proportions at different times.

(II) Kielmeyer's Economy of Extinction

Kielmeyer includes an important qualification in his claim about the distribution of forces in species kinds. He tells us that 'no very marked dissimilarity' was produced, and that excellence of one force in one kind was 'almost always' offset by excellence in another. This brings us to the second purpose of the equilibrium model. According to Kielmeyer, this model is also meant to explain the temporal course of organic nature. Perhaps it is not surprising that Kielmeyer was interested in the course of nature in his inaugural speech, given that, in his earlier lecture notes (1790), he had emphasized the importance of the diachronic character of a genuine natural history when giving his own definition:

> the history of the phenomena yielded by our earth as a whole, according to the concept of natural history, must address not only the question of their present condition, but also that of the states preceding and perhaps succeeding the present one – thus, how it is, how it was, and how it will be.
>
> p. 60

Likewise, in the 1793 Speech, he claims that the 'machine [of the organic world] also appears to be progressing along a path of development that we may best represent with the image of a parabola that never circles in on itself' (p. 47). From this we can at least reject the possibility that Kielmeyer held a cyclical understanding of history in the vein of Mendelssohn:[22] the development of the organic world presented in the 1793 Speech seems clearly linear and suggests an unending development of the organic world. But how does this conjecture about nature's history align with his additional claim, that 'the course of development of the organic realm is founded along with the way that different forces themselves are distributed' (p. 44)?

I described above how Kielmeyer allows for the *possibility* of historical changes in kinds, because postulating a balance of organic forces rather than species populations allows him to posit a persistence of organic force despite (what we might consider quite extreme) phenomenal changes. But what makes such changes 'more than merely possible' in Kielmeyer's account is the additional claim that:

> Despite these compensations, not only to the individual forces decrease at the expense of the others, but also the sum of forces decreases in unknown proportions – and this can be explained neither by the medium in which the animals live nor by other circumstances.
>
> pp. 41–2

And elsewhere:

> since in addition to the greatest possible balancing of all forces with one another, a gradual decrease in the sum of forces in the series of organisations was also noted in the plan of nature, so the emergence of one such over-preponderance is thereby more than merely possible.
>
> p. 44

Kielmeyer had already described this 'balancing' of forces in the laws of distribution that account for the equilibrium of destructive and preserving forces, but here he makes the further claim about a gradual reduction in the total quantity of force attributed to each being as the scale is descended. The most obvious result of this is Kielmeyer's discussion of species extinction, which his conceptual apparatus allows him to frame as consonant with the 'course and persistence' of the organic world.

It bears mentioning that later in his career Kielmeyer had some doubts about the likelihood of extinction events. In a letter Kielmeyer wrote to Windischmann in 1804, he suggested that what could be interpreted as evidence of extinctions (e.g. bones of animals that match no existing varieties) is 'also interpretable as changed *orientation in the formative force*, that entered with the changes of our earth' (p. 67) – so, interpretable as evidence of changes within species in line with the earth's changing geological conditions. In 1804, he does not rule extinction out categorically as a possibility, but he does mention that he thought the relevant phenomena to be 'more likely' (p. 67) to be accounted for by Lamarck's explanation of transformation. However, this seems to be a clear change from his earlier position. In the 1793 Speech, there are good reasons to think extinction is not merely accommodated, but something Kielmeyer intends his theory to explain. He writes that:

> As much as the continued existence of the organic world is in this way founded on those laws, nevertheless it cannot be denied that it is exactly these laws that, on the other hand, also give rise to the question as to whether here and there an over-preponderance in the destructive forces does not actually arise, and whether species whose forces of preservation are inferior must not eventually themselves pass away? Moreover, since, in addition to the greatest possible balancing of all forces with one other, a gradual decrease of the sum of forces in the series of organisations was also noted in the plan of nature, so the emergence of one such over-preponderance is thereby more than merely possible; and, if we now only stick closely to what observation actually tells us, the places in this organic world where over-preponderance emerges – and where it works most forcefully – in fact appear to become somewhat more precisely determined.
>
> This is the condition of many species of mussels, who, by scant compensation in their forces, and by the wealth of forces that nature gave higher animals for [the mussels'] destruction, are placed on such a dangerous peak that one can assume with likelihood that the decline of one or other species is gradually occurring, and has perhaps already often happened.
>
> p. 44

Because Kielmeyer claims that the course of development of the organic realm is founded along with the way that different forces themselves are distributed, and, in addition, that 'a gradual decrease of the sum of forces' can be noted in the comparison of 'higher' to 'lower' animals, an imbalance can occur. The situation that he describes species like mussels to be enduring is that of a gradual decline apparently ordained in the proportions of forces apportioned to them.

In the dynamics described above, the reduction in forces that manifests in mussel species as they decline is not an isolated event. Their decline is an effect of their poverty of forces for repelling the expanding force of another kind.

This can be seen in his discussion of extinction events resulting from human action:

> Humanity also actually displaced other species of animals, transferred them to small areas as strangers while it took the rest of the earth into its possession, indeed it enslaved entire species, and it is more than merely probable that it will compel still many more to completely exit the stage in order to make space for itself, as one organ replacing another in the great machine.
>
> p. 44

What is notable for our purposes in this disturbing extract is that this 'compulsion' to 'completely exit the stage' as humans expand their forces, as well as the image of organ replacement, seem to suggest that Kielmeyer understands the extinction of one type to go along with another species occupying or taking on its prior share of force.

Further, this redistribution of force can also be seen in his discussion of the change in distribution of forces made consciously. He writes that:

> With the rationality that turned up in humanity's organisation, mankind obtained the capacity to freely alter (within certain limits) the ratio of the other forces that it has in common with the other animals. It created microscopes and telescopes for eye and ear, and thereby heightened its sensorial capacity, and who knows whether further similar improvements may not be applied to smell and touch.
>
> p. 44

In Kielmeyer's picture, the tools used by humans form extensions of their organs: this is made possible by their use of reason. Insofar as it is used to further a sensorial capacity, a microscope becomes part of the human organisation. Incidentally, this also suggests that an organic force, on Kielmeyer's account, is not down to some special organic matter, but is instead due to the type of activity or function the force is directed towards – one of the five Kielmeyer outlines. For example, insofar as a technical instrument is used to further an animal's sensibility, it is functionally part of that animal's organisation. In any case, the key issue for our purposes is that taken together, the above suggests not just a simultaneous parallel rearrangement of forces in two species, such as humans increasing their sensory powers (and therefore having a different internal ratio of forces) whilst mussels increase their reproductive power; rather, due to the interrelation between species, one species' increase in forces for its own preservation has the effect of being at the same time an increase in destructive force considered with respect to other species it opposes, i.e. something taken up in one species results in the decline of another. He writes that:

> To increase its capability of movement, it coerced other animals, fire, and wind, to lend theirs to it, and through these changes that it undertakes in the

proportions of its forces, and through its greater capability to endure each of these changes, it obtained with its species a decisive preponderance over most other animal species and their forces of preservation.

p. 44

The 'coercion' of other entities to 'lend' capabilities to humans – taken in combination with his earlier discussion of species of mussels – to me suggests that this should be considered as a redistribution of forces between different kinds. This is important because it means that such instances must be regarded as implying genuine species extinction, rather than (as Coleman argues regarding the 1804 letter[23]) explicable as a reorientation within existing kinds. In Kielmeyer's system, neither species kinds nor individuals must remain constant.

(III) 'How It Will Be'

The analysis I've provided thus far outlines the mechanisms by which Kielmeyer accounts for a historical picture of nature that includes changes like extinction events. In this final section, I want to briefly address the scholarly claim about Kielmeyer's natural history as 'temporalizing the chain of being', and to speculate about the implications of Kielmeyer's account for the future of nature, especially given the way he accounts for extinction events through dynamics internal to the discussion of organic forces.

Bersier and Zammito have argued that Kielmeyer is responsible for a 'temporalisation' or 'animation' of the great chain of being. This mode of argumentation incorporates Kielmeyer into the scheme first discussed in detail by Lovejoy, which holds that a transition from a hierarchical to a temporal chain of being is a key trajectory in the history of thinking about the natural order. Whilst I agree with these scholars that Kielmeyer introduces an important historical path for organic nature into his thinking, I want to note that a model of understanding the organic world on the basis of the distributions of forces between different species does not of itself guarantee a specific form of historical worldview.

What exactly is implied by the claim that Kielmeyer 'temporalizes' the chain of being? To work out the kind of temporalisation at stake in Kielmeyer's account requires a prior decision on what counts as a significant change, and, moreover, what the relevant component units of such a change would be. For example, one model might regard the birth and death of new individuals within a species as not significant, but concern itself with changes at species level, though this is clearly an unstated assumption. The answer to the question about temporalisation also depends on what features that chain is supposed to have. Lovejoy's account of the chain of being involves the fusion of the 'principle of continuity' (that there are no gaps in nature; every kind borders on something else in the continuum) with that of 'plenitude' (whatever potentially exists must actually exist, i.e. there are no unrealized potentials that are genuine potentials). The course of nature Kielmeyer lays down in the speech does not seem to include a commitment to these principles.

Nor does temporalisation in Kielmeyer (at least in his 1793 Speech) seem to progress diachronically up his scale with each higher species emerging in sequence.

Despite this, Kielmeyer does employ a hierarchical ordering in his discussion of the distribution of forces (with more diverse kinds of expression being highest), and he does ascribe significant historical changes to nature, though many details about what this could involve are absent from his exposition here. Species extinctions certainly seem to be a feature of this development, as a consequence of the relative distribution of forces in different kinds. Though humans do seem to be singled out as most capable of influencing the distribution of forces, the extinction of one kind and expansion of another are not restricted to being the effects of humanity alone. 'What so many higher animals are capable of when it comes to some mussel species' he writes, 'humanity and its species are capable of doing with respect to so many other, and even higher animals' (p. 44). In any case, in Kielmeyer's discussion of human activity, using instruments to expand their share of force, we have a clear case where the distribution of forces becomes increasingly concentrated within one species. Because Kielmeyer has given an account of extinction in the foundations of his discussion of organic force, and yet in the speech does not devote attention to an account of how new species might arise, nor any detailed account of the branching of existing species into distinct new ones, the internal trajectory of the organic world could be read as a progressive reduction of species kinds, with remaining kinds holding a greater concentration of force. This seems to locate a tendency internal to the interplay of organic forces with the result that the history of the distribution of organic forces could be one that progressively concentrates forces in fewer beings and species kinds.

This speculation about the result of the initial distribution of forces between species aligns well with claims that Kielmeyer temporalizes the chain of being, insofar as his organic system tends towards concentrating forces in the most divergently differentiated species. This species, unsurprisingly, seems to be humanity, whose initial capacities give it the potential for later technological advances and hence the still greater diversification of its capacities. If this concentration or narrowing were what Kielmeyer had in mind, it would present quite a different picture of the natural world to the emphasis on productivity and proliferation that is often found in the ways Schelling or others working on philosophy of nature around 1800 are understood.

To say whether the 'concentration' I have proposed was what Kielmeyer had in mind or not, or even if it is an unintended consequence of his remarks, is difficult given the multitude of unanswered questions in the 1793 Speech. As I have mentioned, he does not explain whether or how new species or transformations of species could occur (if this narrowing picture were to be avoided), nor give a clear extent of the scope of the extinction described. It is also difficult to come to any conclusive judgements about Kielmeyer's claims here due to his account remaining internal to the sphere of organic forces – let me just mention here a few factors Kielmeyer refers to outside this picture.[24]

First, Kielmeyer doesn't look beyond the general patterns of interrelation of forces to the institution or creation of the system, beyond the claim that proportional distribution grounds the course of nature. To return for a moment to Derham: in his *Physico-Theology*, the equilibrium sustaining the world is instituted by God, with a certain purpose and direction. Whilst in Kielmeyer we see this terminology of equilibrium in the organic world taken up in his 1793 Speech, the broader theological intent is absent, or at least obscure. He discusses the functioning of a system of organic forces (the course and persistence mentioned earlier), but not the institution of this functioning. Kielmeyer's 1790 account of natural history – as requiring investigation not only into the present, but also the past and future – seems compatible with what he later describes as the 'path' of the organic world in his speech. However, in the 1793 Speech – if it is to be considered as a natural history – he does not dwell on the establishment of that world at length, beyond the claim that the path of organic nature is established along with the proportions in which the forces are initially distributed. How those forces came to be in the first place is not part of his inquiry. In this way, Kielmeyer's speculation about the history of nature offers an explanation of the tendency of organic force that is *internal* to the organic forces, seemingly more interested in the functioning of the system given certain epistemological constraints, rather than some kind of external teleology by way of creation.

Second, despite this emphasis on organic force as the causal and explanatory agent in nature's history, it is not clear that Kielmeyer intends the proportions or balances of force sufficient to explain that history by themselves. Some additional explanation is needed to determine the particular forms found in the world; the stipulation of a certain ratio or proportion, as constrained by the possible combinations of force distribution, does not seem to be sufficient to determine the *kind* of expression a force takes.

Third, Kielmeyer doesn't examine the relation of the organic proportions to their underlying, composite conditions – for example, chemical processes. Beyond the claim that the organic forces are likely reducible to them, he keeps causal and relational explanations on the organic register in the speech.

Fourth, Kielmeyer does not look deeply into the effects of forces external to organic force on the path of nature, into questions such as how geological disasters, like earthquakes, disrupt organic systems. As a result, the progression of the organic world can come across, as an internal trajectory. But it is clear that Kielmeyer did not want to deny the possibility that there could be some relation between the organic world and events in the inorganic, such as geological events.[25] For example, he discusses the varied effects of the environment on the development of species and on which internal features are brought out in specific environments. That he provides a set of rules constraining the distribution patterns of organic forces is not to suggest that they are not affected from without, but merely that in the event of some action from outside affecting them, organic forces will consequently have to realign, and that realignment must follow certain rules. As we have seen, Kielmeyer countenances the possibility that previously inorganic force can be co-opted into additional organic power, but this will have the effect of a reduction of organic forces elsewhere or in other species.

The progressive narrowing picture outlined above could be a possible trajectory for Kielmeyer's history of organic nature, if it were not a system that can be affected from without. But it seems clear that geological conditions are meant to have an impact on the organic system. The explanation on the organic level is not total, nor is it meant to be. With his account of proportions of force in equilibrium, we get a certain picture that, all things being equal, it seems the world will follow. But the 1793 Speech is not meant to be a conclusive treatise that gives a complete or final picture of nature; it is rather a sketch that draws out possibilities for further comparisons and research, limiting itself to discussion of organic forces.

Concluding Remarks

Kielmeyer gives a hierarchical explanation of natural forms, in which he establishes a 'scale' of organisms. The relative position on the scale of a given species seems to come down to the extent to which diverse expressions of force are unified in particular kinds of organism – the greater the unified diversity, the higher the organism on the scale. On the other hand, the formation of his key concepts – equilibrium of force, compensation of force and the reduction of all forces to one primary force – suggest a flatter system, in which Kielmeyer is able to countenance changes in his system because equilibrium is located on the level of forces instead of number of individuals. But his additional claim that there is a gradual reduction in the amount of force given to each species as the scale is descended could have the effect that, over long periods of time, more vulnerable species become extinct as the 'higher' species expand their powers.

The equilibrium is such that Kielmeyer considers the system as a whole – regarding the quantity of force that appears in the expressions of organic phenomena – to not be majorly disrupted by even extinction. But his work reveals a somewhat ambivalent attitude to the extinction of species in nature's history. Conflict is viewed as a necessary component of nature's maintenance as well as its development, and the progression of nature seems to involve the acquisition of shares of force from the 'lower' to the 'higher'. For humans to alter not only their own proportions of force, but also their share of total force, even to destroy other species (even if this is due to the use of technology) is part of the 'life' of the world.

Questions can be raised regarding the implications and limits of this picture: this explanation remains on the register of organic forces and the internal dialectic of their interplay, and this trajectory was not necessarily what Kielmeyer envisaged for organic nature. In the 1793 Speech Kielmeyer mainly addresses the history of organic nature in terms of extinction of species and the technological development of humans, rather than the emergence of new forms of life or the transformation of existing forms. This is not to rule out that Kielmeyer considered other historical changes, at this or at other times in his thinking, nor is it to claim that that there would be no possible way for him to accommodate such changes within the framework of the natural economy he outlines in 1793. But it is notable that this is what he chose to highlight in his main published work. Where other accounts of natural economy had often tried to mitigate potential disturbances to species, for example, through their compensation mechanisms, Kielmeyer allows for long-standing changes, extinctions, to take place.

Notes

1 That Kielmeyer himself understood his project in terms of a natural economy is made clear in his 1790–3 lecture notes outlining his idea of a comparative zoology. There, Kielmeyer claims that the goal of comparing features of animals to determine laws of their relationships is the formation of an 'economy of organic nature in general' (*GS* 123).
2 This conversation was precipitated by Timothy Lenoir's *The Strategy of Life* and surrounds (1) the use of teleology in Kielmeyer's account of organisms, (2) Kielmeyer's proximity to Kant in this regard, and (3) the issue of whether Kielmeyer can be aligned to a Kant-influenced 'research program'. See T. Lenoir, *The Strategy of Life: Teleology and Mechanics in Nineteenth Century German Biology* (Dordrecht: Reidel, 1982), and the competing accounts given by John Zammito, Robert Richards, and others.
3 See W. Coleman, 'Limits of the Recapitulation Theory: Carl Friedrich Kielmeyer's Critique of the Presumed Parallelism of Earth History, Ontogeny, and the Present Order of Organisms,' *Isis* 64.3 (1973). I do not think that Kielmeyer introduces this principle in his speech, for the reasons outlined by Sander Gliboff in *H.G. Bronn, Ernst Haeckel, and the Origins of German Darwinism* (Cambridge Massachusetts: MIT Press, 2008), 45.
4 See A. Gambarotto, *Vital Forces, Teleology and Organization: Philosophy of Nature and the Rise of Biology in Germany* (Dordrecht: Springer, 2018), and J. H. Zammito, *The Gestation of German Biology Philosophy and Physiology from Stahl to Schelling* (Chicago: University of Chicago Press, 2017).
5 See R. J. Richards, *The Romantic Conception of Life: Science and Philosophy in the Age of Goethe*. (Chicago: University of Chicago Press, 2002).
6 This claim is made in G. Bersier, 'Visualising Carl Friedrich Kielmeyer's Organic Forces: Goethe's Morphology on the Threshold of Evolution,' in *Monatshefte* 97.1 (2005): 'By transforming the hierarchy of immutable creatures into a dynamic gradation of changing, developing, and self-regulating lifeforms, he lent concrete scientific shape to Herder's vague notion of a gradation of organisations ... In so doing [i.e. in creating a systematic account of the interaction of forces] he animated Charles Bonnet's and Carl Linnaeus' static ladder of creatures'. This claim is repeated in Zammito, *Gestation*, 258.
7 See A. O. Lovejoy, *The Great Chain of Being: A Study of the History of an Idea* (Cambridge, MA.: Harvard University Press, 1971).
8 The use of a natural economy model puts Kielmeyer in line with the projects of Linnaeus, Buffon and others, who used natural economies to conceptualize the interdependent relations of entities in the natural world. In Linnaeus, Buffon and Derham, different approaches are taken to the consideration of balance and imbalance in the natural economy. But in all these varied earlier approaches, lasting imbalances in species kinds are not permitted, and where imbalances in populations are permitted, some mechanism whereby they are shown to be temporary imbalances is described, with compensation mechanisms tending the system back towards its original state. In Kielmeyer, the idea of balance is decoupled from a notion of balance that must preserve species kinds. Where human interference could have previously been put

down to a temporary imbalance, or a contingent adjustment accommodated for by the play in the economy of nature, in Kielmeyer, human interference can have a permanent effect in species extinction.

9   Entry, 'Gleichgewicht', in G. Toepfer (ed.), *Historisches Wörterbuch Der Biologie: Geschichte Und Theorie Der Biologischen Grundbegriffe*, vol. 2. (Darmstadt: Wissenschaftliche Buchgesellschaft, 2011), 99.
10 W. Derham, *Physico and Astro Theology: or, a Demonstration of the Being and Attributes of God*, vol. 1. (London, 1786), 241. Cited in F. N. Egerton, 'Changing Concepts of the Balance of Nature,' in *The Quarterly Review of Biology* 48.2 (1973).
11 See Egerton, 'Changing Concepts', 333.
12 J. Keulartz, 'Rolston,' in W. B. Drees (ed.), *Is Nature Ever Evil?: Religion, Science, and Value* (London: Routledge, 2003), 90.
13 Toepfer, *Historisches Wörterbuch Der Biologie*, vol. 2, 98–112.
14 Briefly, of these forces, three are adopted from Albrecht Haller, and Kielmeyer adds of two further forces, secretion and propulsion.
15 Thanks to Henning Tegtmeyer for making this point in conversation.
16 He writes, 'In the same way that light appears split into different rays, and these rays are mixed in infinitely different proportions, the smallest organ, right up to the most complex and immense machine, is set in motion by a single force, a force perhaps was originally awoken by a light whose daily support it still enjoys' (pp. 18–19).
17 It makes sense that this is then borne out functionally, such that those animals with fewer senses, for example, are greater in number and it would make sense that these kinds would be less able to survive if not multiplied further. One way in which this line of thinking plays out is in Kielmeyer's claim that the number of individuals in a species is offset by complexity of those individuals – a greater number tend to be simpler. This lack of diversity within individuals of a particular kind can be expressed in having fewer senses, or in having a more homogenous build: 'the four thousand and sixty-one muscles that Lyonet found in the field caterpillar are far less diversely disposed and much more similar to each other than the far lower number found in humans' (pp. 36–7).
18 F. W. J. Schelling, *Sämmtliche Werke*, ed. K. F. A. Schelling (Stuttgart) and Augsberg: Cotta, 1856–61), vol. 2, 565.
19 F. W. J. Schelling, *First Outline of a System of the Philosophy of Nature*, trans. Keith R Peterson (Albany: SUNY Press, 2004), 141.
20 Ingrid Schumacher notes that these figures were included in Kielmeyer's philosophical education, see her 'Karl Friedrich Kielmeyer, ein Wegbereiter neuer Ideen. Der Einfluß seiner Methode des Vergleichens auf die Biologie seiner Zeit,' in *Medzinhistorisches Journal* 14 (1979): 81–99.
21 To give a simple example, on this schema, a species possessing a great quantity of reproductive power could be 'offset' by the presence of greater irritability in another species, so a grass, when compared to a hare which grazes on it, might 'compensate' for its lesser capability of reaction and movement by reproducing at a great rate and sustain its number.
22 See M. Erlin, 'Reluctant Modernism: Moses Mendelssohn's Philosophy of History,' in *Journal of the History of Ideas* 63.1 (2002), 83.
23 I also disagree with Coleman's claim that Kielmeyer's later rejection of extinction was on account of his rejection of the 'continuity of creation'. Coleman writes that 'To Kielmeyer extinction would introduce real, not just apparent, multiplicity into organic creation. This hypothesis violated the fundamental, idealized continuity of creation and was therefore to be rejected' (Coleman, 'Limits of the Recapitulation Theory', 342).

Coleman contextualizes Kielmeyer's remarks in relation to Herder, such that 1) no extinctions are held to have happened, and 2) subscribing to progression in species does not imply transformation to a new kind but rather a development to a new stage *internal* to the species. We might think of this as a geological change precipitating the manifestation of a different set of potentials already possessed by a species activated by new geological circumstances (i.e. all these forms are already contained within the egg and the development of that egg into an animal of a particular shape is contingent on its geographical circumstances), or instead, as a gradual transformation effected alongside the change of the earth where the species has an internal history that progresses alongside the earth's history. To show that this is the case, I think, involves proving that these phenomena are not explicable via a transformation within species, in which case, the number of kinds in the organic world would remain constant, or increase, as no lineages would be able to have died out. Of course, Coleman discusses Kielmeyer's later views here, but neither at the earlier or later time in Kielmeyer's thinking is there any reason to support Coleman's idea that Kielmeyer reasoned from an *a priori* subscription to the principle of continuity to facts about the empirical world (in this case, the impossibility of extinction).

24 In addition, even if we remain at the organic level of explanation, the question remains: given the interdependence of species, when one goes extinct, and another takes up its 'share' of forces, an equilibrium of preserving and destroying forces may more or less be maintained. But, what is the effect of this on species that were dependent on the extinct kind for their survival, and, in turn, what are its knock-on effects? Surely, Kielmeyer must have had an explanation, perhaps an explanation that would focus on the realization of different potentials of a kind, or further extinctions – but again, this is speculation.

25 In fact, the idea that these could be conceived of as independent realms is itself problematic, which makes it all the more surprising that Kielmeyer puts this dynamic forward only in the register of organic forces in the speech. Later in his 1804 letter to Windischmann he discusses the connection between the earth's history and that of organisms (although he is sceptical about the possibility of articulating this history in full). He posits the same force at work in the development of species as in individuals, ultimately relating these developments to the magnetism of the earth.

For a discussion of Kielmeyer's changing views on the relation between the history of the earth and the history of organisations, see J. H. Zammito, *The Gestation of German Biology Philosophy and Physiology from Stahl to Schelling* (Chicago: University of Chicago Press, 2017), 261–5.

Chapter 10

RECAPITULATION ALL THE WAY DOWN?
PHILOSOPHICAL ONTOGENY IN KIELMEYER
AND SCHELLING

Iain Hamilton Grant

This essay asks after the philosophical content of Schelling's claim in the 1797 *Weltseele* regarding the epoch-forging character of Kielmeyer's natural history. It will pursue two lines of inquiry. The first is to present Kielmeyer not as the ally or enemy of *Naturphilosophie* amongst the *Naturforscher*[1], but rather as presenting a hypothesis in the 1793 Speech, subsequently developed from the great wheels in the natural machine to the n-dimensional space of developmental natural history entailed by his response to Windischmann's query in 1804. The second is to argue that a version of this hypothesis also underlies his account of natural history in the period contemporary with the encounter with Kielmeyer, as documented in Schelling's essay 'Is a Philosophy of History Possible?', suggesting thereby a reason for Schelling's enthusiasm for the 1793 Speech. Although Schelling keeps overt philosophical company with Kielmeyer only until the 1804 Würzburg lectures, what I will call Schelling's 'epochal ontogeny hypothesis' persists in *The Ages of the World*'s theory of times and the natural-historical concerns of the *Philosophical Introduction to the Philosophy of Mythology* (1847–52), which I examine elsewhere. Finally, that natural history is a Schellingian philosopheme throughout his career partly corroborates Hans Michael Baumgartner's claim that Schelling's is from the beginning a philosophy of history; but it does so on condition that this history is not reducible to the Herderian 'innerness' of the problematics of freedom and evil which Baumgartner argues it is, but acquires its full philosophical significance only once the centrality of natural history to the problematics of history is understood, so that not only humanity but, as Henrik Steffens demanded, the earth and all its productions have both an *inner* and a *natural* history.

If part of my present claim is that Schelling's is a philosophy of history just if it is a *Naturphilosophie* throughout, the question thereby disinterred concerns the ontological, or better, the ontogenetic significance of naturalized history for philosophy and the difference this makes to our understanding of Kielmeyer's specific theory of recapitulation, its scope and its operation.

Natural History and the Genetic Model

Resolving the genetic problem is an Orphic journey beneath appearances, seeking to return the embodied particular, the world-egg, to the phenomenal surface, in the Goethean model.[2] In Herder's reiteration, in seeking to 'understand the nature of things, our urge to know' unsatisfied by revelation and so demanding 'cosmogonies ..., pursues a path into the darkest ages ... to conjecture with probability the beginning of things'. To discover the 'germ cell of life', the 'ultimate source' or 'foundation of the whole', the 'Orphic darkness' must be risked by both poet and *Naturforscher*. The 'philosopher of the infancy of the ages' will then be able to descend through the history 'from which, in the end, everything is derived', and to give 'subsequent shape' to her 'diligently gathered material' by way of this 'genetical explanation'.[3]

What is true for the Ovidian poet is equally so for the Proteus-wrestling natural historians contemporary with Kielmeyer. Herder's Orphic cosmogony makes fellow-travellers of poets, novices, philosophers and natural historians, all pursuing their inquiries by way of what Herder summarizes above under the rubric of the genetic explanation or *method*.[4] Naturalists around the turn of the nineteenth century may be said to have pursued the 'genetic method'[5] when seeking 'to identify the simple type of a given structure' and so forego inquiry into the genesis of 'simple types' and take structures as *given*. Solving the genetic method, that is, is achieved by identifying the 'unit of selection'. It is when natural *history* rather than comparative physiology[6] is pursued that the genetic problem arises. Contrastively to the *method*, the genetic *problem* arises when the emergence of structure *at all* is at issue as opposed to that of types and structures presently scrutable. What makes Kielmeyer's account of natural history philosophically interesting, and so draws Schelling's attention, is that Kielmeyer's 'unit of selection' for the genetic method transforms it into the genetic *problem*. Ironically, just as he extends recapitulation to geogony in an 1804 letter to Windischmann, Schelling, in his contemporaneous lectures in Würzburg, criticizes Kielmeyer's particular claims concerning the process of recapitulative series *in individual beings* as 'contradicted by natural history'.[7]

Apart from the philosophical interest, however, Schelling further holds Kielmeyer likely to have altered the course of contemporaneous natural history. Amongst what may be considered proto-recapitulationists,[8] Herder articulates his own 'genetic method',[9] a precursor of Nietzsche's 'genealogy', as a predominantly semantic archaeology: present conceptions are dependent on as yet unidentified prior conditions. What those conditions are, however, Herder leaves unsettled: underlying uniformity or unfounded relativism? Indeed, Herder's work consists in the construction of antinomies: does history consist of war or work? Is it teleological or diversifying? Progressive or cyclical? Is it the 'innerness' of sensation and cognition or the outward effects of human life that are history's objects?[10] If the distinction between innerness and mere outerness obtains, where is it to be drawn? Or is the innerness proper to cognition and sensation the only innerness that – or wherein – nature knows? In this case, however, is innerness nature's, since if so, then it becomes false that nature has no innerness; or is nature irreversibly innerness's outwardness? It is Herder's address to natural history in the second part of the *Ideen*, however, that concisely exhibits this antinomic character of the genetic problem. Herder first claims that 'Nature needs infinitely many seeds, since in her great path she pursues a thousand goals at once.' Yet shortly thereafter, he argues that:

> it is undeniable that all the variety of living beings on the earth seems to be governed by a certain uniformity in construction and, as it were, by a *primary form* [*Hauptform*] that changes [*wechselt*] into the richest variety.[11]

From the first claim, two problems arise. First, what allows the 'infinitely many' seeds to be grouped to form a more manageable typology? Does Herder intend that the merely 'thousand goals' contrastively facilitates precisely this classification, i.e. that many distinct seeds are yet subject to the same ends? If so, what distinguishes one from another of these 'infinitely many' seeds that contribute to common goals?

Secondly, how in the case of infinitely many seeds is it possible to distinguish between what nature does and what anything else does? If there are infinitely many seeds, no reduction is possible to one species of production as opposed to another, although theoretically an infinite number each of infinite types of production (natural, inner, aesthetic, etc.) might yet remain possible while also allowing for distinctions of type.

Herder's solution, as articulated in the second claim, is that research into the quantitative reducibility of natural phenomena, into the production of species-categories therefore, finds ultimately – should its trajectory be pursued from the infinite to the many and on to a One – a single *Hauptform* or 'uniformity of structure' that, despite the wealth of apparent diversity into which it changes, seems to govern all things. Along with the infinite number of seeds and nature's thousand paths, the *Hauptform* is thus one of Herder's three hypotheses as to the terminus of the genetic method. Other *Naturforschern* propose alternative candidates: Carus's *Urzellen*,[12] Oken's *Urschleim*,[13] Goethe's *Urpflanze*,[14] Ritter's *All-Thier*,[15] amongst others. Each such candidate provokes two questions in turn:

1. how, if there are infinitely many seeds does *one Hautpform* 'diversify' into the 'thousand paths'? and
2. whence – or *when* – this One, if nature is indeed an infinite number of seeds?

In other words, what is the primary of which everything *else* in nature is secondary? Perhaps, to pick up on a suggestion from Carus, the *Grundform* of any organism is the recapitulation (*Wiederholung*) of innumerable recapitulations, these in turn being recapitulations of the 'first germinal seed (*Eikeim*)' consequent upon the 'basal idea, actualized'.[16] The problem remains, however, that if the *Grundform* is the *Grundidee* actualized, i.e. if something has befallen a primal, we are returned to assuming simple types (ideas) or structures (forms), so that once again, the genetic method terminates not in primals but in paradox.

This is indeed an Orphic programme, partly as Wolfram Hogrebe defines the inquiry after meaning as Orphic when the attempt to redeem its object already entails its loss.[17] Yet the Orphic inquirer seeks not only to discover lost epistemic or semantic treasures, but to locate the cosmogonic egg from whence all sprang. Mere revelation, as Herder puts it, is insufficient when a cosmogony is required such as Kielmeyer conceived to be entailed by any comparatist natural *history*,[18] once the basal form assumption has been dropped so that recapitulation does indeed go all the way down, between the large and the small, or between organization at its most extensive (cosmogony) and at its least (time-particles).

In consequence of these contrastive organizational 'time-paths' (*Zeitbahnen*),[19] there arises an additional dimension of this problem; that is, the issue of *when* the cosmogonic egg might be discovered, and in what space it might be found, turn out to be themselves historical questions, and hence an element of the philosophy of natural history into which Kielmeyer *will contribute*, so Schelling, a new epoch.

Kielmeyer and Epochality

Kielmeyer outlines the principal axes of his natural history in the monologue 'Ich [DIE NATUR]' from the first version of the 1793 Speech.[20] It is preceded by a terse presentation of what we may call the ontological commitments of Kielmeyer's comparatism, which it is important to note insofar as these aid in differentiating the account offered in the 1793 Speech from von Baer's recapitulation or the so-called Meckel-Serres Law he is held to foreshadow. Additionally, it is enveloped in a philosophical scheme that positions the functional range of what he calls, following Reinhold, the 'faculty of representation (*Vorstellungsvermögen*)',[21] i.e. subject and phenomenon, with respect to that of *system* and 'nature's greatness'. Hence the 1793 Speech's opening programme:

> If, by the powers of our minds [*Kräfte unseres Geistes*], we separate the phenomena of nature – for us connected in a system by space and time – from their connection, then surely those phenomena that we isolate and subsume under the name 'animate nature' – I mean the organisations of our earth – are those most able to fill us with feelings of nature's greatness among [the phenomena] with which we are closely acquainted. To be sure, no masses, volumes, or distances found here are like those of the skies, by which nature convinces us of its greatness. However, if, when judging the greatness of an object, we can deign to give voice and listen with a little patience to the multiplicity [*Vielheit*], diversity [*Mannigfaltigheit*] and harmony of effects in a small space and short periods of time, then there are things of another kind that speak to us no less forcefully.
>
> p. 30

The opening account of natural phenomena separated from the spatiotemporal connections that they enjoy 'for us' does not contrast the powers of our mind with nature unminded, instead insisting on the availability of two conceptions of natural phenomena which we might call the systemic and the asystemic, respectively. The point of this contrast is to demonstrate that the concept of nature always and necessarily exceeds the system of spatiotemporal phenomena consonant with subjectivity. Not only, then, is nature greater than possible appearances divided by the subject for which they are phenomena, but even this equation, i.e. $\varphi > \left(\dfrac{Ph}{S}\right)$, is itself a product of the 'catalytic' and synthetic powers of our mind.

This address casts light on Kielmeyer's response to Cuvier's inquiry as to his erstwhile professor's view of the 'proposition that external nature ... can be deduced from a priori principles'. Kielmeyer responds that the 'usual sort' of idealist argument that nature is entirely a construction of mind must either deny the existence of objects outside us, treating them instead as sheer representations whose emergence they do not attempt to explain; or must explain emergence on the basis of the nature of our mind and cannot therefore explain the emergence of that nature (pp. 69–71). Attributing the second view (which he rejects as either inconsistent or unfounded)[22] to Fichte or to those Fichteans who maintain that nature is reducible to 'not-I', Kielmeyer is at pains not to deny the active nature of the mind in the process of knowing which philosophy inherits from Kant (pp. 70–1). By contrast, he simultaneously accords greater consistency to the 'real-idealists'[23] and, as per Reinhold's recommendation that systematic philosophy go 'back to first principles', that is, the grounds or 'first, highest and only principles', ultimately supplied by the 'principle of consciousness' as a means to resolve the partiality and idiosyncrasy of philosophical systems, he suggests reversion to 'first ... philosophy' with its original or innate antitheses such as active and passive, subject and object, and so forth. Rather than interpreting this as Reinhold does, that is, as providing the key to systematizing philosophy as 'rigorous science', Kielmeyer follows empirical psychology's reception of Reinhold's philosophy[24] and adopts a view at one and the same time quasi-anthropological, discussing 'humanity's first, innate philosophy' and, importantly, *genetic*. All his criticisms of the 'idealists' focus precisely on the genetic question: if the mind constructs nature, the question is source-directed and therefore invokes the firstness of first philosophy not as Aristotelian ontology, but rather as satisfying the question of beginnings. He thus re-interprets Reinhold *genetically* or, as he puts it in the 1793 Speech, in terms of how such things 'fill time' (p. 30).[25]

Kielmeyer's address to how organisation "fills time" and his reconception of time according, as we will see, to time-paths (*Zeitbahnen*) and time-particles (*Zeitteilchen*), is both philosophical and, as the passage above makes clear, methodological. Methodologically speaking, Kielmeyer recommends that rather than this or that body being taken as the basal referent of comparative analysis, the 'multiplicity, manifoldness and harmony' of the time-filling operations discernible within 'a small space and a short time' provides the measure. The analysis he proposes thus divides space-times by operations whose 'harmony' circumscribe the space and the time filled by the phenomena under examination. This is why, for

Kielmeyer, as for Schelling's '*Scheinprodukte*',[26] phenomena are operations – or, as he exhibits this thesis in 1834:

$$\left\{ \frac{Phenomena}{Operations} \right\}\text{,[27]}$$

There is no question here of de-realizing the less ontically weighted of these elements (*mere* appearance) in relation to the more; rather, since operations exceed the phenomenal framework as the parabola's directrix does its focus, discrete phenomenality or form is a natural accomplishment. It is therefore not those forms themselves but rather the scales of organization across space-times that form the *object* of natural history, while small and short spaces and times or discrete selection forms its method.

Taken together, the two formulae – that nature is more than the phenomena into which subjects divide it, because nature is the excess of operations over the spaces and times in which they are maximally scrutable – give the rationale for Kielmeyer's comparatism, which is between *scales of organization*, such that the 'recapitulation' at issue is not simply between individual and species but extends from the 'developmental history of the earth' (p. 65) to 'the developmental periods of ... the human mind' (p. 66) – in short, throughout all of nature. When, therefore, 'I, Nature' seeks to 'lead you beyond space and time and accompany you along the path where it departs from your system', (p. 46) the point of thus emphasizing nature's disconnectability from local spacetime frameworks is not to draw a naïve realist contrast between nature as it is in itself and nature as it seems to us, as is clear from his critique of subjectivism (p. 72); but instead to picture or conceive the nesting of organization within organization that nature comprises regardless of which scale of organization measures or assesses it.

Hence, the 1793 Speech begins by invoking the abstractive powers of the *Vorstellungsvermögen* which are then demonstrated in operation as the auto-presentation of nature; and ends with the claim that the same forces evinced in the various periods of the development of 'the human mind' constitute a refiguration of those same proportions of forces whose recapitulation is evident in the development of individuals and species (p. 30). Which is nature's 'innerness' here, and which its outward phenomenon? The reality is their mixture, as expressed by the formula by which Kielmeyer places phenomena on the same footing as other natural operations: the forces operating in the one are the same forces as appear in the other.

The importance of what might be called the 'ideal dimension' of nature's susceptibility to presentation consists therefore not in isolating representations from the operations articulating them, but in the demonstration that every presentation of nature whole is also local, while its local character does not subtract from the whole nature so presented. This is because the ideal or the image is itself an instance of the organization whose partial expression it is. The *Vorstellungsvermögen* is not the solution to the genetic problem, but rather a *consequence* of the becoming that, Kielmeyer argues, nature 'is' (pp. 30-1). As Schelling puts the point:

> What really is a developmental stage? It is indicated by a determinate form. But this determinate form is itself simply a phenomenon. The real in which it is grounded is the inner proportion of forces found to be original in every organization.[28]

In other words, natural *history* has no conclusion, no ultimate stage, no perfection. Ideality is not final but provisional. If there is therefore no final form of a nature subject to history, what solution does or can Kielmeyer propose to the genetic problem? If organization is the object of natural history, it is only an object in the epistemic and not in the ontic sense. The methodological point, therefore, that organization consists of *the operations that appearances are*, or morphogenetic events, entails the metaphysical one that only something operative or *wirklich* can fill time. Accordingly and perhaps surprisingly, given his rejection of finality of form, Nature's monologue argues that the fundaments of what 'fills time' are *Zeitteilchen* or 'time-particles' as they develop along 'time-pathways' (*Zeitbahnen*) (p. 30).

Although Nature's monologue contains Kielmeyer's only mention I can locate of his conception of *Zeitteilchen*, the concept of organization as forming periodicities or cycles and thus filling time is neither without precursor nor consequence in the naturalist's slender oeuvre. It remains consonant with his definition of nature in On Natural History (1790):

> Everything emergent or actually appearing to our senses, that is perceived with outer as with inner sense outside and within us, interconnected with space and time and following certain laws, is *nature*.
>
> p. 55

Where the interconnectedness in space and time that becomes a feature of discrete systems in the *Rede* is tied to the determination of appearance as operation that remains a feature of Kielmeyer's thinking even in 1834's essay on the kinetics of plant morphogenesis that seeks not to subjectivize appearance and oppose it to the objectivity of effective action, but to make appearings members of the community of 'actual emergents'.

Taking the time element contributory to the 1793 Speech's natural history in common with the directionality element of the essay on plant morphogenesis alerts us to one further feature of Kielmeyer's formula evident from its form and in the manner in which it is introduced. The key question, Kielmeyer writes in his late plant-essay, is, *without making philosophical assumptions*, what higher and more general classes of $\left\{ \dfrac{Phenomena}{Operations} \right\}$ are presently perceptible in individual plants to which a particular phenomenon may be reduced? The plant-essay's central problem concerns the *antithetical motions* evident in the roots' geophilic or downward development and the stem's upward heliotropism. The formula's immediate context therefore invites us to consider the relative positions of its factors: first, as higher or lower in the hierarchy of logical classes; and there is additionally the question concerning the 'reduction' of phenomena to its fundament or ground. In both cases, the formula's factors simply are by organization higher or lower, phenomenon and ground, in such a manner as to reinforce their separate ontic status. Both accounts therefore seem at odds with the general tendency of Kielmeyer's emergentist nature, since this precisely does not presuppose such a ground or any such reduction; and while there are logically superior and inferior organizations – for example, the organ is higher than the time-particle but lower than the time-pathway – Kielmeyer's entire argument is that there neither is nor will be no settled highest. What therefore can be said about a settled lowest, about a ground or about the time-particle that we have suggested is Kielmeyer's candidate solution to the genetic problem? In considering these issues, which we now pursue via Schelling, it becomes clear that the incontrovertible sublimity of the ceilingless heavens may indeed proffer a symmetrical sublimity in the ultimate groundlessness of this candidate solution.

Schelling on Natural History and Philosophical Ontogeny

Schelling's first encounter with Kielmeyer is as the author of a critical review of his *Ideen zu einer Philosophie der Natur* in 1798.[29] Simultaneously, however, he begins two years of intensive work on the natural historian's ideas before separating considerations of natural history that persist until his last works from direct engagement with Kielmeyer's version of them. There remain strong parallels, between the two, from their questioning of the form of natural history itself to their curiously parallel kinetic theories of dimensionalization in the 1830s and 1840s.[30] It is, however, in a short essay from 1798 that Schelling first sets Kielmeyer's ideas for a natural history in a philosophical context.

Schelling's 'Is a Philosophy of History Possible?' (1798) inverts the question proper to the genetic problem as to the source from which structure developed (if structure is the sort of thing that develops) and asks instead whether a rational or systemic account of history is possible at all. Schelling conspicuously poses this problem through the lenses of natural history, rather than addressing it, for example, in the Herderian framework of 'innerness' historically available to him as a thesis concerning the proper content of history.

Just as Kielmeyer's 1793 Speech is self-consciously (pp. 30, 45–6) bookended by the question of mind or ideality, 'History' begins at the ideal end of the problem. In neither case is ideality considered, as Herder might, as 'innerness', for reasons most directly explained through Whitehead's claim that a 'private psychological field', where 'private' is understood not as individual but as disconnected from any and all additional fields, and so 'merely the event considered from its own standpoint',[31] i.e. the pursuit of discreteness at the cost of any possible comparatism such as Kielmeyer, and Schelling after him, advocate. Instead, ideality plays the same role in history as do other appearances, articulating only an alteration in the proportion of forces involved in its production. This integration is not the 'materiality of the Idea', which thesis Schelling repudiates in his 1794 *Timaeusschrift*, i.e. the reduction of ideality to 'simply located bits of matter',[32] because it is not *the object* but rather *an iteration* of historical process, a cosmogonic feature.

The first and 'broadest concept' Schelling offers of history asserts, first, that the proper object of *die Geschichte* (history) is *das Geschenen* (what has happened) rather than *das Bleibende* (what remains), and secondly that this object is nature, as the 'totality [*Inbegriff*] of everything that has happened', i.e. as belonging to the *past*. In the form here presented, this thesis, a version of which will become an element of Schelling's *Weltalter*, has the consequence that time is parcelled out between nature and its present and future successors. To it Schelling adds a contrastive thesis concerning history, namely, that it is progressive, which progressivity filters out 'regularly recurring' natural phenomena as properly historical. This discrimination is based on a concept of nature as obedient to *a priori* and thus necessary law. As a result, the developmental interiority of post-natural history is progressively insulated from a nature to be consigned to a prehistorical past or, as Schelling formulates the point,

> ... what must be judged a priori and occurs according to necessary law is no object of history; conversely, what is an object of history cannot be accounted a priori.[33]

This 'broadest concept of history' remains, accordingly, insufficiently broad to supply a natural history, and to that extent corresponds to the history consequent on Herder's innerness thesis. Accordingly, should a Herderian *Hauptform* obtain, Schelling's argument entails, then as the *a priori* form of nature's 'infinitely many seeds', it too, by definition, fails to supply a natural history not only because it cannot account for the emergence of innerness but also because it sets a developmental ceiling on the 'infinitely many seeds' since, on the assumption of nature's segregable pastness, their totality is necessarily finite.

If by contrast nature's segregable pastness is abandoned, the question as to a finite 'primal form' resulting from it invokes Kant's distinction between *Naturbeschreibung* and *Naturgeschichte*,[34] that is, between a nature susceptible of description as opposed to the nature as 'a system of operations' entailed by a naturally historicizing nature. Schelling seeks to divest a concept of a system of natural history of the last vestiges of the descriptivism that reduces 'natural history' to a 'faithful copy' in favour of a philosophy of properly historical natural history. While not wholly eliminating the *picturing* that descriptivism offers, Schelling stakes the possibility of a system of natural history in place of the genetic method's hypothesis of its 'basal type',[35] in order to demonstrate that systematicity and historicity are not mutually exclusive. He offers two hypotheses by which 'a natural history in the strict sense' is possible, his account of the consequences of which I will quote at length. Either

1. 'all individual organizations instantiate different developmental stages of one and the same organization, in which case the archetype of all of them must lie in the midst of our current system of nature'; *or*

2. 'the current state of nature is maximally differentiated from its original state [...which] assume[s] an original variability in organic nature'.

In the former case, the 'system' of natural history would

> ... think nature in its freedom as it evolves along all the possible paths that accord with an original organization (which for that reason cannot exist anywhere now) just as now one force, at the cost of suppressing others, has here a lower and there a higher intensity, so that an equilibrium is reached between these same organic forces; this would still have the advantage over the second idea of providing for a reduction in the outward variations amongst the earth's creatures as regards the number, magnitude, structure and function of their organs to an original, inner variation in the interrelations of forces (of which these were but the outward phenomenon). In the latter case, we would not of course need to establish a common archetype for all organizations; but we would certainly need common archetypes for the many now distinct kinds that owe their origin to a gradual deviation from the original, produced by various natural influences.[36]

Schelling's argument is not that one case should prove the 'faithful copy' at the expense of the other, but rather to accept (a) that such archetypes as might obtain must do so 'in die Mitte' of a system of natural history and (b) that it is only when they do that 'nature's history' obtains. In other words, the phenomenon of nature as a whole is not given at the outset but arises amongst the system of nature, or has that system as its medium. Hence the question becomes the *naturphilosophische* one of how nature gives rise to its self-exhibition, whole.

If, however, this whole of nature should arise and so makes a 'copy', an exhibition, *a* conception or 'one phenomenon' of 'nature's history',[37] in what respect does this remain *historical*? Schelling's stipulation that history obtains 'where there is an ideal and an infinite multiplicity of deviations from it in individuals, and where there is nevertheless complete *congruence* with it on the whole', seems at first sight to eliminate rather than to secure history in the interests of the species of a-priorism Kielmeyer agrees with Cuvier is 'delusional'. Yet the 'infinite multiplicity of deviations ... in individuals' makes natural production profligate rather than merely primal, which natural production must include that of the ideal from which these productions deviate since, as arising 'in the midst' or the medium,[38] they will not be antecedent but consequent upon the nature or 'emergence', the ideal of which they exhibit. Where 'the whole' is thus at any given moment always less than it will be and so 'historical' precisely because, even as the summative focus or being of that system, like the 'derived absoluteness' whose consequent nature is such that 'no kind of combination can transform ... into what is by nature original',[39] qua consequent, it has history as its medium and is therefore *amongst* those deviations to which, in summing them, it contributes. We may therefore derive from this a Schellingian formula of recapitulation: *x recapitulates y just when both x and y are maximally different from their origin*, since any emergence recapitulates emergence per se. History, accordingly, is always *becoming* what has happened and is in this respect its most encompassing definition. Rather, therefore, than eliminating history, what is eliminated is anything original, or, in other words, Being.

Conclusion: From the Biogenetic Law to Philosophical Ontogeny

> I have often had the idea of a cosmogony proceeding by analogy with an organism's development.
>
> Kielmeyer (*GS*, 57–8)

For *Naturforscher* and *Naturphilosophen* alike, therefore, history supplants ontology's priority. A philosophy of history is possible therefore when the systematic ambitions of the former are realized in the medium of the latter. The net result, which Schelling will continue to interrogate throughout the near two decades of work on *The Ages of the World*, is a philosophical ontogeny. This arises not on the basis of the genetic method and its uncovering of a basal type, but rather with the extension of the systematic interrelation of forces or operations (*Wirkungen*) that appear not differently to what they are 'at root', but insofar as they emerge and so, according to Kielmeyer's definition, add to nature's paths even when simply duplicating them.

If the history essay, published in July 1798, some five months after the *Weltseele*'s appearance, contains an elaboration of Schelling's lavish encomium to Kielmeyer therein, it becomes apparent what the new epoch Kielmeyer will be held to introduce might consist in. Drawing on the genealogy of epistemic states that opens the *Weltalter*, the prophetic form, referencing future states, becomes immediately significant. Rather than pursuing an Orphic egg lying behind or before nature's appearance, the structure of nature as a whole will be a natural historical event. A being is historical, Schelling accordingly claims, just when it has no *given* history but 'brings it out of itself, produces it'.[40]

We are thus thrown back to the two hypotheses the history essay offers: arising in the medium of the history it presents, entails that the natural history of nature's exhibition is 'maximally differentiated from its original state'. The new epoch Kielmeyer's natural history introduces into natural history is precisely that which divides a natural history premised on assumptions concerning a basal type and pursuing some variant of the genetic method to disclose it, from one whose very historicity makes such unity the late and provisional accomplishment of a freely productive nature whose history is being made. If nature *is* historical, that is, it knows no bottoming out whereby the genetic method will discover its basal type unless history were, paradoxically, a late acquisition of the egg at its base.

Finally, Schelling's 'epochality' additionally marks a philosophical engagement with the temporal frame within which Kielmeyer articulates his history of a nature that comprises 'everything emergent'. As such, the medium of natural history cannot consistently be this or that entity but consists of *Zeitbahnen* and systems of *operation* or *appearings*, of cosmogenesis at one extreme of these systems and pathways, and time-particles at the other. Departing therefore from what Kielmeyer tells us but drawing on this reconstruction of the philosophical impetus his natural history provides Schelling, in what might time-particles consist? The parts of time are not segregable units, from which no time, as Zeno's paradoxes make plain, may be constructed, but assume particle form in respect of *time, just when* nature is 'maximally differentiated', as Schelling put it, i.e. when new epochs are not rare but ubiquitous. Kielmeyer's time-particles thus compose the elements of a philosophical ontogeny as first philosophy, following the 'elimination [of the ...] concept of original being' from nature philosophically conceived in consequence of the advent of a natural history wherein 'history must itself become nature through and though if it was to assert itself as nature, that is, in every aspect of its being.'[41]

This would be why we are mistaken in considering Kielmeyer's recapitulation reducibly a forerunner of the biogenetic law, namely, that the development of individuals recapitulates speciation; in Schelling's presentation, that same work cannot support so regional an account of recapitulation, which instead, in Kielmeyer's stated views, spans not simply the relation of individual to species, but encompasses all appearances or operations from cosmogony to the noogony that has as its consequence the becoming of a nature that becomes self-portraying thereby. The naturalism to which Schelling responds in Kielmeyer does not seek to differentiate nature from history, to isolate candidates for innerness from uniform outerness, nor to isolate the archetype of all things from which everything sprang, since the nature Kielmeyer presents as consisting of 'everything emergent' was not once creative only to become a creature of law faithfully self-copying to eternity, but the maximal differentiation that is nature and that obtains just when anything emerges rather than nothing.

Notes

1 The standard strategy in reading Kielmeyer is to demonstrate him philosophically inclined or disinclined toward *Naturphilosophie*. Compare R. J. Richards, *The Romantic Conception of Life: Science and Philosophy in the Age of Goethe* (Chicago: University of Chicago Press, 2002) and T. Bach, *Biologie und Philosophie bei C. F. Kielmeyer und F. W. J. Schelling* (Stuttgart: Frommann-Holzboog, 2001).

2 'The metaphysics of phenomena proceeds out of the largest and from the smallest (only visualized for humans by artificial means); adapted to our senses, the particular lies in the middle, but my heartfelt blessings to those gifted ones who bear those regions up to me.' J. W. Goethe, *Schriften zur Naturwissenschaft* (Stuttgart: Reclam, 1977), 49; my translation. See also I. H. Grant, *Philosophies of Nature After Schelling* (London: Continuum, 2006), Chapter Four. Goethe's natural scientific works have recently received partial confirmation – see O. L. Müller, *Mehr Licht! Goethe mit Newton im Streit um die Farben* (Berlin: Fischer, 2015).

3 The quotations here are from J. G. Herder's 'Fragments of a Treatise on the Ode' and 'Essay on a History of Lyric Poetry', found in E. A. Menze and K. Menges (eds), *Johann*

Gottfried Herder. Selected Early Works 1764–1767 (Pennsylvania: Pennsylvania State University Press, 1992), 69, 36, 71, 50, 70.
4   Many have made the case that Romantic *Naturforschung* issues first from Herder. See, for example, Richards, *The Romantic Conception of Life*, 245: 'Kielmeyer obviously built upon notions of the historical development of species inaugurated by Herder and Blumenbach'. Roger Ayrault goes further, claiming that German Romanticism as a whole is no more than a development of ideas first presented by Herder. See R. Ayrault, *La genèse du romantisme allemand* (4 vols., Paris: Aubier, 1961–76).
5   F. G. J. Henle, *Handbuch der rationellen Pathologie*, 2 vols. (Braunschweig: Vieweg, 1846–53), 1:25, cited in E. Clarke and L. S. Jacyna, *Nineteenth-Century Origins of Neuroscientific Concepts* (Berkeley: University of California Press, 1987), 21.
6   In this respect, my argument diverges from Thomas Bach's assertion (in *Biologie und Philosophie*, 277) that 'Schelling primarily or in the first place too over the conception of a comparative physiology from Kielmeyer' just as if there is something else that Schelling took from Kielmeyer.
7   F. W. J. Schelling, *Sämmtliche Werke*, ed. K. F. A. Schelling (Stuttgart-Augsburg: Cotta, 1856–61), VI:454. Hereafter: *SW*.
8   Gould (*Ontogeny and Phylogeny* [Harvard: Harvard University Press, 1977], 49) notes that Goethe, Herder and Kielmeyer were all candidate Ur-recapitulationists, and Ayrault (*La genèse du romantisme allemande*) makes the case that Herder is the source not just of so-called romantic biology but of romanticism in general.
9   Herder, *Selected Early Works*, 70: 'Just as the seed contains, concealed within itself, the entire plant with all its parts, the source of an art contains within itself the entire nature of its product'. See M. N. Forster's discussion of the 'genetic method' in *After Herder. Philosophy of Language in the German Tradition* (Oxford: Oxford University Press, 2010), 36–41.
10 For his concept of 'innerness', see J. G. Herder, *Philosophical Writings*, ed. M. Forster (Cambridge: Cambridge University Press, 2002), 204–10.
11 J. G. Herder, *Ideen zur Philosophie der Geschichte der Menschheit*, ed. Heinrich Luden (Leipzig: Hartnoch, 1841), II/ii: 38, 49.
12 Directly responsive to the subject matter particular to Clarke and Jacyna's (*Nineteenth-Century Origins*) account of the 'genetic method', C. G. Carus's 'On the Unconscious Life of the Soul' (in *Geheimnisvoll am lichten Tag*, ed. Hans Kern [Leipzig: Reclam, 1944], 49) traces the origins of the animal nervous system to a 'psychical principle' contained in the semi-fluid masses of the *Urzellen* from which 'all true animals' issue.
13 L. Oken, *Gesammelte Werke*, vol.2: *Lehrbuch der Naturphilosophie*, ed. Thomas Bach et al. (Weimar: Böhlaus, 2007), §900: 'Every Organic has issued out of mucus [*Urschleim*, i.e. protoplasm], is naught but mucus under different forms.'
14 J. W. Goethe, *Die Metamorphose der Pflanzen* (in E. Trunz [ed.], *Goethes Werke. Hamburger Ausgabe* [München: Beck, 1975], vol. XIII, 25) establishes his method as seeking the 'pure form', which he elsewhere calls the '*Urphänomen*'(primal phenomenon), that both 'embraces' and is 'identical with all cases' (Goethe, *Schriften zur Naturwissenschaft*, 42).
15 J. W. Ritter, *Beweis, dass ein beständiger Galvanismus den Lebensprozess in dem Thierreich begleite* (Weimar: Industrie-Comptoire, 1798), 171: 'Where is there a sun, where is there an atom that is not at the same time a part, that would not belong to this *organic universal* . . ., the great *Universal Animal* of *Nature*?'
16 'On the Essence of the Primary Formative Process in the Human Organism', in Carus, *Geheimnisvoll am lichten Tag*, 35. The passage reads in German: 'Wer . . . nur das recht

gefaßt hat, wie alle Grundform des Organismus auf unzählbarer Wiederholung der *einen* Grundform ruht, und wie jegliche Zelle der Wiederholung des ersten Eikeims ist und wie sie selbst eben dadurch immer wierder die Grundidee verwirklicht oder eben dadurch in sich eigenlebendig ist, der betrachtet nun schon mit ganz anderen Augen das aus all diesen Wiederholungen sich erbauende Ganze.'

17 W. Hogrebe, *Metaphysik und Mantik. Die Deutungsnatur des Menschen (Système orphique de Iéna)* (Frankfurt: Suhrkamp, 1992), 9. It is interesting to compare the pattern of Hogrebe's Orphism with Marcel Detienne's in *Les dieux d'Orphée* (Paris: Gallimard, 2007, 25). In Hogrebe, since the search for an object must uncover its prior absence, that recovery repeats its loss. The pattern is therefore loss-revelation-loss. In Detienne's Orphism, however, drawing on the Orphic bone tablets discovered in 1951 at Olbia, Sardinia finds 'life-death-life' to be the pattern associated with 'truth'.

18 See Kielmeyer's claim: 'I have often had the idea of a cosmogony proceeding by analogy with an organism's development.' (*GS*, 57–8)

19 Kielmeyer's term '*Zeitbahn*' articulates the vector character of a temporal extension, or of 'appearing' as a 'System von Wirkungen', systems of operation or effect (p. 30).

20 This passage, which is translated above, is not included in K. T. Kanz's facsimile edition of the 1793 Speech (Marburg an der Lahn: Basiliken Press, 1993) but, as found in the Württembergischen Landesbibliothek, Stuttgart, Cod. Med. Et phys. Fol.38a. is printed in *GS*, 63–7., where it is introduced as an 'excerpt from the live first version of the *Rede* delivered in 1793'.

21 K. L. Reinhold, *Versuch einer neuen Theorie des menschlichen Vorstellungsvermögens* (Jena: Widtmann and Mauke, 1789).

22 Inconsistent in that, to explain this construction, such idealists must either assume an external determination of the active I, contradicting its stipulated nature or 'fantasize' some third thing apart from I and not-I from whence the mind is 'fertilized' into action; unfounded therefore in that this last leaves it entirely unexplained how the antithesis subject-object arises in the first place. (pp. 71–2)

23 Schelling advocates Real-Idealism in, e.g. *SW* IV:89.

24 See R.-P. Horstmann, 'Maimon's criticism of Reinhold's *Satz des Bewusstseins*' in L. W. Beck (ed.), *Proceedings of the Third International Kant-Congress* (Dordrecht: Reidel, 1972), 330–8, where he notes the importance of Reinhold's philosophy to empirical psychology as evident from the fact that Maimon's address to it was published in Karl-Philipp Moritz's *Magazin für Erfahrungsseelenkunde*, and by works such as C. E. Schmid's *Empirische Psychologie* or L. H. Jakob's *Grundriß der Erfahrungs-Seelenlehre* (both 1791), where the way the traditional elements of empirical psychology are combined with Kantian and Reinholdian elements may be studied in detail.

25 I prefer 'fill' to 'occupy' since the former emphasizes the dynamic or operational character of temporal articulation while the latter places what occupies squarely in that and only that space-time.

26 Schelling, *SW* III:16. See Grant, *Philosophies of Nature after Schelling*, Chapter Three, for an extended discussion of phenomenal production.

27 C.f.. C.F. Kielmeyer and G. Jäger (eds) *Amtlicher Bericht über die Versammlung deutscher Naturforscher und Ärzte zu Stuttgart*, (Stuttgart: Mezler, 1835), 59: $\left\{ \dfrac{Erscheinungen}{Wirkungen} \right\}$ or: 'Die bei allen einzelnen Pflanzen wahrnehmbare Richtung ihrer Wurzeln nach unten, erdwärts, und die Richtung der Stämme nach oben, himmelwärts'.

28 Schelling, *SW* III:54.

29 Schelling's vexation with the review is evident in a letter to his father of 19 May 1798: 'You may already have gathered what might be expected from Tübingen from the review of my *Ideas* . . . . The author is Kielmeyer who, lacking better company, joins those wretched individuals whose closest friend he is.' F. W. J. Schelling, *Historisch-kritische Ausgabe*, vol. III/1, ed. Irmgard Möller and Walter Schieche (Stuttgart-Bad Canstatt: Frommann-Holzboog, 2001), 162.
30 Compare Kielmeyer's 1834 essay on plant kinetics with Schelling's *Dimensionenlehre* in the *Darstellung des Naturprozesses* of 1844 and his lecture on the movement of animals in *Darstellung der reinrationalen Philosophie* (1847–52).
31 A. N. Whitehead, *Science and the Modern World* (Cambridge: Cambridge University Press, 1926), 186.
32 Ibid., 72.
33 Schelling, *SW* I:467
34 I. Kant, *Critique of Judgment* trans. Werner S. Pluhar (Indianapolis: Hackett, 1987), 315; and 'On the Different Races of Human Beings' and *On the Use of Teleological Principles in Philosophy* in I. Kant, *Anthropology, History, Education,* ed. Günter Zöller and Robert B. Louden, (Cambridge: Cambridge University Press, 2007): 89, 197–8.
35 See Clarke and Jacyna, *Nineteenth-Century Origins*.
36 Schelling, *SW* I:468–9. For the clearly Kielmeyer-inspired phrasing, compare the title of Kielmeyer's 1793 Speech.
37 Ibid., *SW* I:470
38 Schelling similarly positions being 'in the middle' in the course of his *Stuttgarter Privatvorlesungen* (ibid., *SW* VII:436): 'Our being is sheer medium, an instrument for us [*Unser Seyn ist nur Mittel, Werkzeug für uns selbst*].'
39 According to the law of antecedence in the *Freiheitschrift*, (ibid., *SW* VII:347, 340, respectively).
40 Ibid., *SW* I:472
41 See H. Steffens' *Lebenserrinerungen* (Jena: Eugen Diedrichs, 1908), 176–7.

Chapter 11

KIELMEYER AND THE CYBERNETICS OF THE ORGANIC WORLD

Andrea Gambarotto

(I) Introduction

In a lecture course held at the *Muséum* in 1832, Georges Cuvier, who had been Kielmeyer's student in Stuttgart from 1784 to 1788, referred to his former mentor as the 'father of *Naturphilosophie* who did not want to recognize his daughter'.[1] The question of how to interpret this aphorism forms the subject matter of the present chapter.[2]

In what sense can Kielmeyer be understood as the father of *Naturphilosophie*? What theoretical innovation did he introduce that contributed to generate the bizarre and historically polymorphic phenomenon we call Romantic *Naturphilosophie*? There are two main possible answers to this question. I will refer to these as the *developmental* and the *physiological* accounts of Kielmeyer's legacy.

The developmental account has been the mainstream option in English-speaking literature. This reading has its theoretical ancestor in the work of Stephen Jay Gould and was developed as a historical narrative by Robert J. Richards. In his seminal *Ontogeny and Phylogeny*[3], which relies on Coleman's earlier historical study[4], Gould identified Kielmeyer as one of the fathers of recapitulation theory, according to which 'ontogeny recapitulates phylogeny,' – a principle which was to have extraordinary implications for the theory of species evolution. In this context, the importance of Kielmeyer is generally attributed to a passage that appears at the end of the 1793 speech, where, after a series of detailed analyses of the distribution of 'organic forces' in nature, Kielmeyer submits that 'these laws, by which the forces are distributed to different organisations, are also precisely those by which the distribution of forces to the different individuals of each species occurred ... to one and the same individual in its various periods of development.' (p. 42).

This statement can be understood as inaugurating the general *naturphilosophisch* attitude to the relation between microcosm and macrocosm, epitomized by the opening statement of Oken's *Abriß des Systems der Biologie* (1805): 'What else is the animal kingdom than the anatomy of humans, the macrozoon of microzoons?'[5].

In this perspective, human anatomy gathers in one individual organism all the organs expressed across the whole of the animal kingdom, and is thus to be regarded as the general model for the overall organization of the organic world.[6] Based on this ontological assumption, it made sense to think that in the course of its development from amorphous substance to formed adult, the embryo passes through all previous degrees of nature's ladder. Accordingly, the adult forms of 'lower' classes should ultimately be understood as arrested developments of 'higher' ones. This view was expressed by the so-called Meckel-Serres 'law of parallelism' between the individual development of an organism and the universal organization of the animal kingdom.

These claims constitute pre-evolutionary remarks on the correspondence between the ontogenetic stages of a given individual and the series of different organic types. Translated into the evolutionary framework by Ernst Haeckel, the hierarchy of the animal kingdom emphasized by *Naturphilosophen* was re-interpreted as the phylogenetic process leading from ancient to more recent configurations of animal morphology, resulting in the idea according to which, in the course of its developmental process, an embryo would pass through all adult forms of its 'less evolved' ancestors. In the late twentieth century, as a result of Gould's work, this view has been reframed as a problem concerning the difference in developmental timing, or heterochrony, in different species[7], which issued in a renewed emphasis on the relevance of developmental processes in understanding evolutionary dynamics – an aspect that had been largely overlooked with the establishment of the Modern Evolutionary Synthesis from the 1940s onwards. In their landmark paper on the subject, Gould and Lewontin[8] borrowed none other than the German notion of a structural *Bauplan* to stress the fact that development makes for a fundamental *explanans* of evolvability. In the course of the last decade, these ideas have witnessed increasing attention in philosophy of biology, due to the rise of Evo-Devo and the quest for an Extended Evolutionary Synthesis.[9] This historical moment has made it fashionable to read Romantic *Naturphilosophie* – and Kielmeyer as its putative father – as the champions of developmentalist approaches to evolutionary theory.[10]

This is a very interesting, and indeed important approach, but, in my view, it ultimately fails to grasp the real importance of Kielmeyer's address. The above-quoted passage from Kielmeyer's *Rede* comes at the end of a series of analyses that constitute the core object of the speech itself, and it appears more like an afterthought rather than a theory in the proper sense of the term. On the other hand, a consistent theory had been developed in the earlier part of the speech, and this concerns the distribution of 'organic forces' in the natural world.

This theory has motivated the second account of Kielmeyer's legacy, which originates, especially, from the German-speaking literature. Key here were the re-edition of Kielmeyer's 1793 speech by Kai Torsten Kanz[11] and also the corresponding conference proceedings[12] aimed at rehabilitating Kielmeyer as a key figure for German biology and Romantic philosophy of nature. An excellent monograph by Thomas Bach (*Biologie und Philosophie bei C. F. Kielmeyer und F. W. J. Schelling*) contributed to this movement by investigating the pivotal role played by Kielmeyer in the development of Schelling's early *Naturphilosophie*. However, this line of inquiry has been less visible in the English-speaking literature, although much more attention has been granted to it in recent years.[13]

In what follows, I develop some of the philosophical implications that can be drawn from this physiology-centred approach to Kielmeyer's legacy. I do so by using twentieth-century cybernetics as a theoretical mirror-image for the philosophy of nature of German Idealism. Although this might appear a strange move, I intend to argue that there is much to be gained by reading some of the metaphysics of German Idealism through the theoretical lenses of authors like Gregory Bateson and Francisco Varela.

Section II sets out the conceptual foundations for my argument. I attempt to argue that the Idealist notion of *Geist* can be profitably re-interpreted in the light of Bateson's 'ecology of mind'. On this line of reasoning, I submit that the call for an account of 'nature as subject,' shared by both Schelling and Hegel, implies a radical rethinking of 'subjectivity' as a universal process, one which generates and grounds 'subjective', transcendental understanding as a derivative form. This process is what German Idealists referred to as 'the absolute' or the 'subject-object', which I take to be just another name for the 'systemic nature of mind'. Section III provides what I define as a 'cybernetic reading of Kantian teleology'. I argue that we can divide contemporary interpretations of Kant's 'critique of teleological judgment' into three main categories: *heuristic*, *pluralist* and *radical* approaches. I take the radical approach to be the best way to make sense of Kant's controversial legacy. I argue that this reading is largely convergent with the way Schelling and Hegel approached it in the direct aftermath of the third *Critique* and I go on to contend that twentieth-century cybernetics can provide the theoretical framework to revitalize this approach today. Section IV provides the historical background necessary to understand Kielmeyer's speech on organic forces in its context, focusing especially on the history of German physiology in the second half of the eighteenth century. Section V then proposes a cybernetic interpretation of Kielmeyer's 1793 speech, showing how this interpretation applies to Schelling's early philosophy of nature as well. Section VI tries to draw some general conclusions from my analysis.

(II) Biological Organization and the Ecology of Mind

In addition to studying for two years with Blumenbach in Göttingen from 1786 to 1788, Kielmeyer spent most of his life at the *Karlsschule* in Stuttgart, first as a student then as a professor, although he also served as a professor of chemistry and botany in Tübingen from 1796 to 1804. Most notably, the speech that appeared in print in 1793 had a huge, if underrated impact, on the development of Romantic philosophy of nature, inaugurating an entirely 'new era of natural history' in Schelling's words. What did this 'new era of natural history' consist of? This is, of course, to ask once more: What does it mean to understand Kielmeyer as the father of Romantic *Naturphilosophie*? My answer to these questions has to do with the development of what might be defined as a 'cybernetics of the organic world'. To explain what I mean, we need to go back to the ideas of Gregory Bateson from which I draw this idea.

In a passage from his *Steps to an Ecology of Mind*, Bateson quotes Alfred Russel Wallace, the younger colleague of Darwin who had discovered natural selection at the same time as his more illustrious peer. In an essay sent to Darwin from Indonesia, Wallace writes:

> The action of this principle [the struggle for existence] is exactly like that of the steam engine, which checks and corrects any irregularities almost before they become evident; and in like manner no unbalanced deficiency in the animal kingdom can ever reach any conspicuous magnitude, because it would make itself felt at the very first step, by rendering existence difficult and extinction almost sure to follow.[14]

Bateson brilliantly reads this passage as evidence that Wallace was proposing 'the first cybernetic model' dealing with 'self-corrective systems' which are 'always *conservative* of something'.[15] In what follows, I submit that, in fact, Kielmeyer should be granted that honour.

A reasonable starting point for my argument would be a clear definition of a cybernetic model. Bateson defines 'cybernetics, or information theory, or communication theory, or systems theory' as the science that emerged after the Second World War to deal 'especially with the problem of what kind of thing is an organized system'.[16] Such a system is defined by the following features:

1. The system shall operate with and upon *differences*.
2. The system shall consist of closed loops or networks of pathways along which differences and transforms of differences shall be transmitted.
3. Many events within the system shall be energized by the respondent part rather than by impact from the triggering part.
4. The system shall show self-correctiveness in the direction of homeostasis and/or in the direction of runaway.[17]

These features define biological organization as a self-regulated cybernetic system, which according to Bateson contribute to generate the 'minimal characteristics of mind.' Accordingly, an 'ecology of mind' should be roughly understood as the science dealing with the 'laws of order, negative entropy, and information'[18] that contribute to generating what we understand as mindedness.

This approach implies a radical rethinking of mindedness as an ecological, instead of subjective, phenomenon, one in which nature is not understood as a material object opposed to a Kantian transcendental subject, but rather as the structural process through which every form subjectivity is generated to begin with, in dialectical coupling with its respective biosphere. In this sense, when Bateson argues that phenomena such as 'mental processes, ideas, communication, organization, differentiation, pattern, and so on, are matters of form rather than substance'[19], he is not far from Hegel's call for a conception of 'the absolute' not only as substance, but just as decisively as subject. Of course, 'subjectivity' here is far from the Cartesian/transcendental understanding of the term, which Hegel had rejected, from his Jena writings onwards, as 'philosophy of reflection'; rather, it designates the hallmark of every self-organizing, autonomous system.

To elaborate on this idea, Bateson makes use of the Jungian-Gnostic terms, *pleroma* and *creatura*: 'The pleroma is the world in which events are caused by forces and impacts and in which there is no "distinctions." Or, as I would say, no "differences." In the creatura, effects are brought about precisely by difference. In fact, this is the same old dichotomy between mind and substance'.[20] We could translate this dichotomy into the German terms, *Natur* and *Geist*. What German Idealists usually referred to by that obscure term, 'the absolute', is nothing more than 'what is the case' in the broadest possible sense of the term. And 'what is the case' in the broadest possible sense of the term is 'nature', conceived as the absolute identity of its material, inert and mechanical component with its formal, living and purposive one. These two sides can be separated for scientific purposes (as the German Idealists did, distinguishing a philosophy of nature from the philosophy of mind), but to conceive this separation as ontologically consistent is the quintessential trademark of abstract thinking.

Iain Hamilton Grant has emphasised the Schellingian idea of philosophy as the 'natural history of our mind', whose main remit is to move beyond the 'two-world metaphysics' bequeathed by Kant with his dualism between 'concepts of nature' and 'concepts of freedom', towards the 'absolute identity of mind and nature'.[21] Located precisely at the crossroad of these two worlds, biological systems constitute the privileged ground to define the fundamental features of an ecology of mind. For example, the fundamental move of the *Weltseele*, whose subtitle runs significantly, 'a hypothesis of higher physics for the explanation of the universal organism', is in fact to overturn the classic empiricist tenet (not dissimilar in nature from today's reductionism) that the mechanism we find at play in the world of inorganic bodies should be held as the fundamental explanatory principle for nature as a whole. Schelling argues, to the contrary, that this mechanical principle of the inorganic should be sublated into a higher, more comprehensive principle – that of the organism. In the *Erster Entwurf eines System der Naturphilosophie*, published a year later, Schelling defines mechanism, chemical affinity and teleology as different 'potencies' that characterize different levels of nature. At lower levels – those corresponding to the world of the pleroma-elementary compounds are extrinsic to one another and interact only through mechanical relations; at higher levels, magnetism and chemical affinity testify to the existence of other intrinsic interactions, whose character is determined by the relation among the terms in play. In the realm of living organisms, the whole thoroughly determines the structure and function of single parts. A similar framework is outlined in the *Objectivity* section of Hegel's *Science of Logic*, which is explicitly divided into three parts entitled 'mechanism', 'chemism' and 'teleology', and which then turn into 'the idea of life' and ultimately 'the idea of cognition'. The general idea behind this schema is that organization should be understood as a higher explanatory principle with respect to mechanism, since without that principle it is ultimately impossible to make sense of phenomena such as life and cognition.[22]

As Grant points out, 'it had been an integral element of Schelling's nature-philosophy since *On the World Soul* that the perimeter dividing organic from 'anorgic' nature be eliminated as naturalistically untenable and philosophically vicious, in order that organization become not an exception to a mechanistic natural order, but rather the *principle* of nature itself'.[23] Re-interpreted in terms of organization, cybernetic self-regulation and homeostasis, Schelling's naturalization of the platonic notion of 'world soul' might appear less outlandish, as just another name for 'the *systemic* nature of mind'[24]: the 'idea of an *organizing*, self-systematizing *principle*. Perhaps this has is what the ancients wanted to hint at by the *soul of the world*'.[25]

Bateson specifies Lamarck and Wallace, along with Samuel Butler, as historical milestones on the way to his 'ecology of mind'. My main point here is that this sort of ecological approach to life and mind can be found in Romantic *Naturphilosophie* and that Kielmeyer, as Schelling himself argued, was the first to provide the theoretical framework for this approach. It is in this sense, then, that I aim to answer the question concerning Kielmeyer's role as a father to *Naturphilosophie*, as inaugurating an entirely new era of natural history. If the 1793 speech functions as the birth certificate of Romantic *Naturphilosophie*, this is, first and foremost, a matter of the development of a view of the organic world as a self-organizing system, which in turn grounds a conception of *Geist* as an ecological phenomenon.

(III) A Cybernetic Reading of Kantian Teleology

To understand the genesis of the above, we need to go back to German Idealism's grappling with the theoretical issues left open by Kant's critical project, and notably by his understanding of teleology as presented in the *Critique of the Power of Judgment*.

The theoretical tension that emerge from Kant's treatment of teleology can be exemplified by two passages from two key paragraphs of the third *Critique*, §65 and §75. The former is important because it contains Kant's most explicit statement on the distinction between organisms and machines, focused precisely on the notion of self-organization; the latter includes the famous passage on the 'Newton of the grass-blade' in which, after emphasizing the difference between organisms and machines, Kant nonetheless maintains that it is reasonably impossible to account for purposive self-organization naturalistically.

In §65, Kant argues that:

> an organized being is thus not a mere machine, for that has only a *motive* force, while the organized being possesses in itself a *formative* force (*Bildungskraft*), and indeed one that it communicates to the matter, which does not have it (it organizes the latter): thus it has a self-propagating formative power, which cannot be explained through the capacity for movement alone (that is, mechanism).[26]

In this passage, he seems to lean towards a non-reductionist approach to biological organization, which, he argues, cannot be explained through mechanism alone. To explain it, we need to introduce a different principle, a peculiar 'formative force' capable of accounting for the self-organizing features of biological systems.

In §75, however, Kant retreats from this position with his famous declaration:

> we can boldly say that it would be absurd for humans even to make such an attempt or to hope that there may yet arise a Newton who could make comprehensible even the generation of a blade of grass according to natural laws that no design (*Absicht*) has ordered.[27]

The notion of *Absicht* (translated above as 'design', but which, in a more general sense, means 'intention') stands in plain contradiction to what Kant had stressed in §65, when putting the emphasis on the intrinsically *self*-organizing features of 'organized beings' as opposed to mere machines. If §65 insists on the principle of organization in living organisms as fundamentally different from the extrinsic, intentional project we find at play in the production of artefacts, §75 marks a step back, in which we have no other conceptual means to make sense of organization than the reference to intentional design.

The contradiction between these two instances of Kant's theory are embodied in the so-called 'antinomy of teleological judgment', which consists in the unsolvable opposition of two maxims. Thesis: 'all generation of material things and their forms must be judged as possible in accordance with merely mechanical laws.' Antithesis: 'Some products of material nature cannot be judged as possible according to merely mechanical laws (judging them requires an entirely different law of causality, namely that of final causes)'.[28] The well-known solution Kant proposes to this dilemma is the distinction between 'determinant' and 'reflective' judgement. The former refers to the 'constitutive' properties of an object, the latter to the 'regulative' way in which our cognitive faculty makes sense of the object. According to Kant, we must consider living organisms *as if* they were the products of intentionally acting causes, while nonetheless dealing with them within a mechanistic framework of explanation. The ultimate meaning of Kant's response and its implications for our understanding of teleology is controversial and still largely open to debate.

I argue that three fundamental answers are live options in the literature: 1) a *heuristic approach*, according to which Kant's treatment of teleology as mere heuristics is coherent, and, in fact, prefigures, contemporary evolutionary approaches that reconcile mechanism and teleology by explaining design as a statistical result of natural selection; 2) a *pluralist approach* that relies on the Kantian distinction of two criteria for a natural purpose to frame this solution as the template for an integration between adaptationism and developmentalism; and 3) a *radical approach* that stresses the contradiction inherent to Kant's troubled relation with teleology and ultimately believes that we rather need to move beyond Kant if we are to provide a theoretical account for the self-organizing features a play in biological systems. This latter position endorses a 'radical' position within

Kant scholarship, by arguing that, if we are to get anything done with Kantian teleology we need ultimately to do so *in opposition to*, or at least *beyond* Kant. I take this radical approach to be the correct reading of Kant's legacy for the problem of teleology – one which is largely convergent with the way Schelling and Hegel interpreted the issue in the direct aftermath of Kant's third *Critique*.

1. The *heuristic* approach has a long history in Kant scholarship. For example, in his study of teleology and mechanics in nineteenth-century German biology, Lenoir sees Kant as the pioneer of a naturalized research programme that recognizes teleology as a fundamental hallmark of living systems, but 'without regrets'[29], i.e. without the philosophical consequences inherent to vitalist positions postulating 'substantial forms' or 'entelechies' as the sources of vital organization. As some critics have remarked, this approach grants a role to teleology in biological explanation, but ultimately does so 'with regrets'[30], i.e. explaining teleology away as a result of mechanical causes instead of giving teleology its due as a fundamental feature of biological systems. Angela Breitenbach represents a paradigmatic example of this approach, when she submits that 'the crucial contribution of the Kantian account is to argue both that teleology plays an important heuristic role in the search for causal explanations of nature and that it is for us an inevitable analogical perspective on living beings. The Kantian perspective, I shall argue, is not only compatible with the modern life sciences but can advance the debate about teleology in biology precisely because it does not interpret teleology naturalistically'.[31] This is not far from what we currently witness in mainstream biological theory, the so-called 'Modern Evolutionary Synthesis', which considers teleology as a post-factum result of natural selection, thus apparently reconciling mechanism and teleology. In this perspective, although teleological features are recognized as undeniable properties of organismal form and function, they are ultimately reduced to the design operated by the 'blind watchmaker' of natural selection.

2. The *pluralist* approach relies on Kant's explicit distinction between a 'design' concept and a 'self-organization' concept of teleology as presented of § 65 of the *Critique of the Power of Judgment*, where Kant famously defines two criteria for something to be a 'natural purpose': (1) It must be a *purpose*, thereby showing an *analogy* to design-like artefacts, in the sense that 'its parts (as far as their existence and form are concerned) [must be] possible only through their relation to the whole'.[32] This feature implies that the 'idea of the whole' precedes the generation of an 'organized being', as in the case of artefacts; (2) It must be *natural*, that is display specific self-organizing features that mark a *difference* from artefacts: for example, a clock does not generate another clock, nor does it self-repair after damage. The former has been defined in the literature as the 'design criterion' and corresponds to what Kant refers to as 'extrinsic purposiveness' (*äußere Zweckmäßigkeit*); the latter can be defined as the 'self-organization' criterion and corresponds to what Kant refers to as 'intrinsic purposiveness' (*innere Zweckmäßigkeit*). Based on these two criteria, Philippe

Huneman[33], has recently framed Kant as the theoretical template for a synthesis between adaptationism and developmentalism in the context of current debates over the prospects for an 'Extended Evolutionary Synthesis'. In so doing, he interprets 'self-organization' as coextensive with development, and, in fact, turns it into the 'epigeneticity' criterion, even if epigenesis is only one form, albeit an exemplary one, of the self-organizing aspects we find at play in biological systems, such as self-regulation, homeostasis or autonomous agency.

3. The *radical* approach puts the emphasis on the second feature of the Kantian concept of purposiveness, one which has been largely underrepresented both in Kant scholarship and mainstream biological theory. Most notably, John Zammito has argued that 'if biology must conceptualize self-organization as actual in the world, Kant's regulative/constitutive distinction is pointless in practice and the (naturalist) philosophy of biology has urgent work to undertake for which Kant turns out not to be very helpful'.[34] In a similar vein, Chilean biologist Francisco Varela and co-author Andreas Weber 'boldly advance the conclusion that, after two centuries, we can move *beyond* the unstable position set out by Kant in the *Critique of Judgement*, and therefore provide a fresh re-understanding of natural purpose and living individuality'.[35] The reason for this instability can ultimately be ascribed to Kant's life-long troubles with teleology and, notably, to his notion of a 'technique of nature'. Although Kant played a fundamental role in setting the stage for a theory of teleology as autonomous self-organization by conceptually defining the very notion of intrinsic purposiveness in opposition to Wolffian accounts of teleology as God's intention, he was ultimately unable to move beyond the conceptual ground of early-modern natural philosophy.[36] The only way he had to reconcile this early-modern concept of teleology with the transcendental limits set for human knowledge was thus to distinguish between constitutive and regulative uses of teleological principles. This distinction is ultimately to be conceived as the result of a capital *failure* on Kant's part to account for the very phenomenon he made theoretically pertinent.

As Hegel did not fail to insist, defining intrinsic purposiveness as the hallmark of biological organization was 'one of Kant's greatest services to philosophy', which 'opened up the concept of *life*'[37], yet, owing to his adherence to a Wolffian understanding of teleology, Kant ultimately failed to capitalize on this fundamental intuition.[38] For German Idealists like Schelling and Hegel, Kant's concept of intrinsic purposiveness was therefore unfinished business, a good intuition to be consistently developed on its own terms. A large part of their philosophy of nature – especially of organic nature, which evidently plays a pivotal role for both of them – can thus be understood as a response to the issues left open by Kant's 'unstable position'. As previously mentioned, I interpret this response as a move towards a cybernetic, i.e. self-organizing, homeostatic, concept of purposiveness.

The use of these terms requires further clarification. The notion of 'cybernetics' makes us think of servo-mechanisms at work in machines like a thermostat, and this has indeed (at least partially) been the point for first-order cybernetics ever since its inception.[39] Rosenbleuth, Wiener and Bigelow stressed the inherent homogeneity of self-regulating servo-mechanisms in living organisms and artificial machines and, in fact, Wiener went on to define cybernetics as the science of 'control and communication in the animal and in the machine'.[40] Teleology, in this sense, defines any kind as behaviour regulated by negative feedback, no matter whether it occurs in organisms or artefacts. Studies in the history of cybernetics agree that the landmark paper by Rosenbleuth, Wiener and Bigelow can be considered as a first attempt at 'mechanizing teleology', yet I submit that this was also the first attempt made in the twentieth century to take intrinsic teleology seriously. As Geof Bowker puts it, 'Rosenbleuth, Wiener and Bigelow attempted to demonstrate that purposive behaviour was a function of negative feedback – and thus not confined to beings with intentions'.[41] The key point of negative feedback as a theoretical tool was thus to conceive of the intrinsic purposiveness at work in organized systems beyond the model of intention. So, despite apparently perpetuating the machine metaphor by other means, the concept of feedback can, instead, be understood as the first contribution to its ultimate deconstruction. This is especially true for the so-called 'second-order cybernetics', which explicitly parted ways with early engineering approaches, but it is also true for other founding fathers such as Ashby, Bateson and McCulloch.[42]

Firmer Northrop devoted his doctoral dissertation to 'The Problem of Organization in Biology,' and so exhibited 'a clear understanding of how the increase in differentiation accompanying the growth of organisms was compatible with the Second Law of Thermodynamics'.[43] The very fact of order-creation, i.e. organization, in the face of the perturbations of the environment (which tends towards disaggregation and disorder) necessarily implies a form of purposive activity – one that is strictly connected to the notion of information.

Pitts' and McCulloch's interest in machines where 'messages, coding, stored programmes, and feedback loops played a role'[44] was primarily aimed at understanding complex phenomena of perception and cognition. Their research question was: 'How would one construct a machine to recognize some of the patterns a human recognizes, or, more generally, to carry out some class of functions characteristic of human mental activities?'[45] The question concerning the cognition of *Gestalten* was a key issue, and, in this respect, the 1958 paper by Pitts, McCulloch and Maturana, entitled 'What the Frog's Eye Tells the Frog Brain', can roughly be understood as the origin of cognitive science.

The assimilation of organismal cognition with the machine was further criticized by James Gibson, founder of ecological psychology, with his distinction between 'sensation', conceived as a passive reception of stimuli, and 'perception', understood as active information-seeking and cognitively selective activity. For Gibson, 'the organism is not a machine processing all inputs from the environment, but an active, selective creature that needs and wants'. Despite this criticism, however, the notions of 'information and feedback-loop play an important role in Gibson's views', although the 'formal-logical computer model does not'.[46]

The point of this brief reconstruction is to stress that there is much more to cybernetics than servo-mechanisms. That 'more' concerns compelling questions on how complex systems are capable of organizing themselves and maintain their organization in spite of external perturbations. This phenomenon is not reduced to ontogenesis, but extends rather to a broad range of phenomena including homeostasis, self-regulation, perception, cognition and agency.

More fundamentally, my point is that these phenomena have a lot to do with what Kant, in his *Critique of the Power of Judgment*, defined as 'intrinsic purposiveness'. And so, I want to address the following question: can Romanic philosophy of nature be understood as developing the Kantian notion of intrinsic purposiveness in terms of self-organization, information and control? In the following sections, I contend that this is precisely the case. When Schelling endeavoured to develop a philosophy of nature that would complete what Kant had left unfinished, he looked to Kielmeyer as the template for his attempt. Schelling makes explicit reference to Kielmeyer in his concluding remarks to the *Weltseele*, where he argues that the organic world is characterized by a 'sequence of functions (*Stufenfolge der Funktionen*)'. These functions refer to the so-called 'organic forces' of 'sensibility', 'irritability' and 'reproduction', which played a fundamental role in the history of German physiology in the second half of the eighteenth century. In Schelling's view, the importance of Kielmeyer lay in establishing the laws through which the distribution of those forces in the organic world is regulated, so as to maintain an overall balance among them. Through these laws, we are led to the idea that '*all these functions are just ramifications of one and the same force*, and that the *only natural principle we have to admit as the cause of life manifests itself in them as its single manifestation*'.[47] This unifying force is what regulates the self-maintenance of nature's 'universal organism'. According to Schelling, this idea is confirmed by 'the *progressive development of organic forces* in the *series* of organizations, with regard to which I refer the reader to the lecture of Professor *Kielmeyer* that already appeared on this subject in 1793, a lecture from which a new era of natural history in the future is without doubt to be expected'.[48]

It is impossible to fully understand the breakthrough Schelling is attributing to Kielmeyer without a general sense of the status of physiology and natural history in late eighteenth-century Germany, especially with regard to the above notion of 'vital' or 'organic' force. And so it is to these subjects that I turn in the following section.

(IV) Organic Forces as a Research Programme

My aim in this section is to synthetically reconstruct the tradition of German physiology in which Kielmeyer's notion of 'organic force' needs to be located, if we are to appreciate its theoretical cogency. I submit that the fundamental breakthrough of Kielmeyer's 1793 speech was to apply the concept of 'organic' or 'vital' force he inherited from the German physiological tradition to the broader context of natural history.

Kielmeyer attempted to demonstrate on the basis of inductive evidence that particular laws regulate the distribution of 'organic forces' (what we would define as 'vital functions') among different classes of organisms. Despite its limited length, this document is extremely relevant for two reasons: (1) because it outlines, for the first time, a general theory of the organic world, which could be considered as the earliest systematic programme for biology as a unified science in Germany; (2) because of the extraordinary impact this programme had on the development of Romantic *Naturphilosophie*.

The only possible point of departure for this reconstruction is Albrecht von Haller's seminal lecture on the sensible and irritable parts of the body, delivered in 1752 and published in 1753, with the title *De Partibus Corposis Humani Sensilibus et Irritabilibus*. This essay, whose importance can hardly be overestimated, was the result of a series of experiments conducted between 1750 and 1751 in Haller's laboratory at the University of Göttingen, in collaboration with his assistant Johann Georg Zimmermann. These experiments were aimed to test the presence of peculiar 'vital properties' in specific parts of the body of several model organisms. These vital properties were 'sensibility,' inherent to nerves, which allows for the reception and transmission of stimuli, and 'irritability,' inherent to muscular fibre, which accounts for contraction upon stimulation. These experiments were incredibly cruel by today's standards, involving the vivisection of the model organisms and the repeated laceration, burning or picking of their organs in order to test the presence of these properties.

Borrowing an expression from Nicholas Jardine[49], Haller's notion of 'sensibility' and 'irritability' opened up an entirely new 'scene of inquiry', and served as a point of departure for physiological research throughout the entire second half of the eighteenth century. In France, such notions contributed to as the development of the so-called 'vitalist school of Montpellier'[50] and, in Germany, they laid the foundations for what has been dubbed the 'Göttingen school'.[51] The central figure of the latter movement was Johann Friedrich Blumenbach, who, after Haller, can be understood as the most important physiologist to teach in Germany during the second half of the eighteenth century. His notion of 'formative drive' (*Bildungstrieb*) played an especially important role as a theoretical 'attractor' for both natural history and philosophy in the following decades.

The importance of this concept and, with it, Blumenbach's parting of ways with Haller can be measured only in relation to the background provided by another key figure in eighteenth-century physiology, Caspar Friedrich Wolff. Wolff was senior to Blumenbach, but younger than Haller, and, after defending a doctoral thesis entitled *Theoria generationis* in 1759, he sent Haller a copy of his dissertation. This gesture, followed by Haller's critical review on the *Göttingische Anzeigen von gelehrten Sachen*, sparked the most important debate in the history of early-modern embryology, which has been excellently reconstructed by Shirley Roe.[52]

Ironically enough, even if Haller's 'vital properties' played a fundamental role in generating eighteenth-century 'vitalism', Haller himself was a micro-mechanist in physiology and held preformationist views in embryology. He believed that the specific properties inherent to nerves and fibres were the result of God's original design, which was already at work from the earliest stages of embryonic life. The rival theory of epigenesis, on the other hand, argued that organic structures were progressively formed in development, as the process through which an embryo passes from being an amorphous substance into a fully formed organism. This was precisely the point made, with painstaking experimental detail, in Wolff's original dissertation, as well as in the revised version, published in German five years later, as *Theorie von der Generation*, which includes a number of responses to Haller's criticisms.

The fundamental contribution of Wolff's theory was the first detailed analysis of development in all its intermediate stages, showing how embryonic forms progressively emerge from one another and thereby invalidate the existence of pre-formed structures with compelling empirical evidence. Yet, if structures were not pre-formed, how to explain the causal force generating organic forms in development? To answer this question, Wolff postulated an 'essential force' (*vis essentialis*) responsible for organic development. The precise epistemological status of such force, however, remained highly problematic. In his mind, this was not even properly a 'force', but rather a general property inherent to organic substances that attracts similar and rejects dissimilar matter and that was supposed to explain the generative process.

The importance of Wolff for the maturation of Blumenbach's views on the nature of development can hardly be questioned. In 1789, in the latter part of his career, after leaving Germany to join the St Petersburg Academy of Science, Wolff instituted a competition for the best essay on the notion of 'nutritive force', which Blumenbach won. The tone of the text is largely accommodating to Wolff's views and there are no signs of a real controversy, yet Blumenbach's famous *Bildungstrieb* essay tells a different story. In the first edition of the essay, Blumenbach argues that the phenomenon of 'generation' testifies to the existence of a 'drive (*Trieb*) (or tendency [*Tendenz*] or aim [*Bestreben*]) which is entirely different from the general properties of the bodies, as well as from the other peculiar forces of organized bodies in particular), which is the cause of all generation, nutrition and reproduction, and which, to avoid all misunderstanding and distinguish it from the other natural forces, I define as *nisus formativus*'.[53] This drive should not be confused 'with the *vis essentialis*, or even with chemical fermentation ... or else merely mechanical forces'.[54] Wolff's *vis essentialis* was a chemical force: the process of generation was driven by chemical attraction and repulsion and controlled by the mechanical properties of the parts of an organism, by means of a chain reaction in which each organ secreted another based on its mechanical-chemical nature. Blumenbach, on the other hand, postulated a vital agent which implied a specific goal-directed action. This 'formative drive' is a goal-directed tendency toward organization, and so it reveals a considerable gap between organic and inorganic nature.

Armed with the notion of *Bildungstrieb*, Blumenbach went on to synthesize Wolff's epigenetic views with Haller's 'vital properties'. In his *Institutiones physiologicae* (first edition 1787), Blumenbach postulated 'vital forces' responsible for three main orders of phenomena: (1) generation, i.e. formation and development (formative drive), (2) motion and contraction (irritability), (3) sensation (sensibility).[55] This unified physiological framework, including three main 'vital forces' as fundamental explanatory principles of organic activity, constitutes the theoretical background out of which Kielmeyer's address on 'organic forces' can be understood. If Blumenbach's contribution to this debate was the integration of Haller's physiological notions of sensibility and irritability into a theoretical framework that included epigenesis and self-organization via the notion of *Bildungstrieb*, the most important aspect of Kielmeyer's contribution was a shift in the application of this conceptual framework from the domain of physiology to that of natural history, with the attempt to explain the distribution of organic forces not in the individual body, but in organic nature as a whole.

This 'physiological systematics'[56], as Stéphane Schmitt defined it, has been correctly identified as the first research programme for a general biology by Kanz and Bach, and I submit that the ultimate theoretical stakes of this programme can be best understood if we read it through the lenses of cybernetic self-regulation not dissimilar in nature from the one Gregory Bateson attributed to Wallace. The idea of self-organization inherent to Blumenbach's *Bildungstrieb*, as the fundamental hallmark of the organic, was translated by Kielmeyer onto the entire organic world, considered as a self-organizing and self-maintaining universal organism. The 'laws of compensation' formulated in the speech can thus be interpreted as the teleological 'control mechanisms' through which 'the great machine of the organic world' maintains its own homeostasis.

(V) The Great Machine of the Organic World

In Section III, I stressed that there is more to cybernetics than mere negative feedbacks. Self-regulating machines, such as the classical thermostat, provide us with models to understand autonomous agency and the complex systemic features of biological system, which are not the result of a designer's intention, but of specific self-organizing laws. In this section, I contend that Kielmeyer's programme for a 'physics of the animal kingdom', as presented both in his lecture notes and in the 1793 Speech, constitute an attempt to formulate precisely such laws.

As specified in the title of his address, those laws concern the 'relations between organic forces in the series of different organisations'. The term 'organisations' is employed as a synonym for 'organisms', which are linked 'in a system of effects ... [as]the life of the great machine of the organic world' in such a way 'that, in our manner of speaking, each is cause and effect of the other', while 'also appears to be progressing along a path of development that we may best represent with the image of a parabola that never closes in on itself' (p. 30). 'Our' way of speaking is manifestly Kant's definition of a natural purpose, and the parabola image is frequently used by Kielmeyer as a symbol for open-ended transformation, in opposition to the recursive processes (represented by the circle). The same image is used in the lecture notes Kielmeyer used for his course of comparative zoology, held in Stuttgart between 1790 and 1793, where Kielmeyer defines his research programme as a 'physics of the animal kingdom' (*Physik der Tierreichs*), which should address:

> (a) The number of organs in the machine of the animal kingdom or the number of animals in general, and the laws according to which these forms are divided into different groups, causes, consequences, or purposes of such division; (b) The relative position of the organs in the machine of the animal kingdom, or the division of the animal kingdom into groups on the earth (geography) according to their different characters, laws of diversity between different groups, causes and effects of such diversity; (c) The related formation of organs in the animal kingdom, gradation of animals and relationships in their formation in general and according to their group, laws, causes and effects of this gradation; (d) The changes that have occurred to the animal kingdom and its groups on earth. History of the development of the animal kingdom in relation to the eras of the Earth ... symbolized by the parabola; (e) Changes repeatedly suffered by the animal kingdom and its groups, the life of the machine of the organic world or its physiology, symbolized by the circle.
>
> GS, 28–9

This research programme was never converted into an extensive piece of writing, but the 1793 speech constitutes a clear outline of its fundamental features. Like Blumenbach, Kielmeyer considered natural purposiveness to be a feature of the organic world: 'suppose that nature had no intent in artificially placing phenomena one after another and next to each other in time, that those effects and consequences were not ends that [nature] wanted to achieve', nonetheless, we still 'have to admit that in most cases this chain of cause and effect looks like a chain of means and end to us, and we even find it conducive to our reason to accept such a chain' (p. 31). Lenoir's *The Strategy of Life* reads this as an example of Kielmeyer's Kantianism and, in fact, the passage lends itself to such interpretation. Kielmeyer employs the term *Absicht* to define purposiveness in nature, and yet he argues that it is necessary to assume such a purposiveness to be at play in biological organization. This does not seem far from Kant's notion of a 'technique of nature' which we must assume as a 'regulative principle', since it is impossible to explain even the generation of a blade of grass without referring to the notion of intention. On his part, Richards' *Romantic Conception of Life* stresses that these statements should rather be read as counterfactuals, precisely in contrast to Kant's regulative reading: even if biological organization were not purposeful (which is evidently the case), it would still be epistemologically reasonable to assume that it is.

I generally agree with Richards here. If the alleged convergence between Kant and Blumenbach on the use of teleological principles should be understood as a 'historical misunderstanding'[57], since Blumenbach ignored the transcendental intricacies of Kant's distinction between 'determinant' and 'reflecting' judgement, the case of Kielmeyer moves such a line of argument even further: he was perfectly aware of Kant's transcendental philosophy and explicitly disagreed with it. In the famous letter sent to Cuvier, asking him to elucidate the new philosophical trends in Germany, Kielmeyer responds that Kant's philosophy exhibits 'significant weaknesses', and that his distinction between subjective-formal and objective-material components of knowledge is 'uncertain and unproven' (p. 74).

The main questions of the address thus concern purposiveness as an objective feature of organic nature, and, namely, the 'teleological mechanisms' through which the 'universal organism' of nature maintains its own homeostasis. On the basis of Blumenbach's schema, Kielmeyer distinguishes five organic forces in nature: (1) *Sensibility*, or the ability of the nerves to retain representation; (2) *Irritability*, or the ability of muscles and other organs to respond to stimulation through contraction; (3) *Reproductive force*, or the ability of an organization to restore injured parts and produce a new ones of like kind; (4) *Secretive force*, or the ability to deliver different fluids to the right places; and (5) *Propulsive force*, or the ability to move fluids through vessels (p. 32). Only the first three play a role in the address, while the latter two are completely disregarded.

Sensibility is the first organic force Kielmeyer considers. Empirical evidence, he contends, shows that the ability to receive different sensations decreases across organisms, moving from higher mammals to lower classes. Among quadrupeds, birds, snakes and fishes, all sense organs are perfect; among insects, however, the auditory and olfactory organs are largely missing. The brain and most of the nervous system found in other animals is not present in worms, for which a single organ collects all sensory stimuli. In plants, this receptivity to external impression is only obscurely present. From these observations, Kielmeyer establishes that '*the manifoldness of possible sensations in the series of organisations decreases as the fluency and refinement of the remaining sensations increases*' (p. 35).

A similar argumentative pattern is pursued for irritability and reproduction. Kielmeyer points out that the animal classes defined by high irritability include pre-eminently cold-blooded animals and insects, which are precisely those characterized by a lower degree of sensibility. Based on his observations, Kielmeyer argues that '*estimated according to the permanence of its expressions, irritability increases as . . . the diversity of sensations decreases.*' (p. 37) Observations of reproduction suggest that 'higher' organisms generate a relatively small number of offspring, and in a longer amount of time, than 'lower' ones, in such a way that '*the reproductive force – evaluated according to the number of new individuals produced in a specific location – increases as the size of the producing individuals, or more universally, the individuals produced as they appear after birth decreases*' (p. 39).

Kielmeyer generalizes these observations, contending that 'sensitivity is gradually displaced by excitability and the reproductive force, and, in the end, irritability too gives way to the latter: the more the one is increased, the less there is the other, and sensibility and reproductive force tolerate one another least. Further, the more one of these forces is cultivated on one side, the more it is neglected on the other' (p. 41). In general, then, the faculties of sensation, widely developed in higher animals, gradually decrease in lower classes and are endowed with great irritability. In the lowest classes (insects, worms), irritability is gradually replaced with reproduction.

This brief reconstruction of Kielmeyer's speech will suffice for the point I want to make. For, bearing in mind the above, I am now able to apply Bateson's definition of a cybernetic model, outlined in Section II, to Kielmeyer's approach to organic forces. In the picture just provided:

1. The organic world is a system that operates upon differences, since the attribution of vital functions to each part of the 'great machine' is regulated by differential relations among those parts.
2. Such relations establish a closed network through which differences are transmitted by feedback loops, since, whenever an organic force increases in one area of the system, it must necessarily decrease in another.
3. These events within the system are caused by the respondent part rather than by impact from the triggering part. In this sense, as Kant would say, each part is at the same time cause and effect of itself.
4. As a result of this, the 'universal organism' of nature operates self-correction towards homeostasis.

The same model is employed in Schelling's *Erster Entwurf*, where he argues that different animal classes are nothing other than the result of different relations among organic forces. They differ from each other not primarily in regard to their material composition, but in terms of the relative proportion of vital functions they display. Like Kielmeyer, Schelling develops a comparative physiology of organic functions whose aim is to establish the various degrees and proportions of each function in the system of nature.

Each species of organism is defined not by its external form, but by the proportion of organic functions active within it. Every organic being is permeated by all three organic forces, but, for example, plants possess a prevalence of reproduction, whereas their sensibility is close to zero. Among mammals, on the other hand, sensibility is dominant, but they produce few offspring and the reproductive force is so narrow that they retain only the capacity to reproduce through growth, assimilation and maintenance. Accordingly, the variety present in organic nature results from variation in the proportion of these functions. They stand in inverse relationship to one another such that, as the one increases in intensity, the other must diminish, and conversely as the one diminishes, the other must increase.

This explains why lower classes are endowed with high reproductive force, while totally or mostly lacking all other vital functions; classes higher in the series of organisms display less reproductive force but higher irritability, which allows them to escape predators; finally, the highest classes, such as mammals, are characterized by sensibility and higher cognitive faculties, which culminate in the perfection of human beings. This is the case because the distribution of functions in the organic world is regulated by a network of inverse relationships, which guarantees 'the continuity of *organic functions* as principle of organization'.[58] Thus 'as *irritability increases in the phenomenon, sensibility must decrease*, and inversely *in the proportion that sensibility increases, irritability must decrease in the phenomenon*'.[59]

As I have already argued elsewhere, the authors of the so-called 'Göttingen school', to which Kielmeyer belongs, played a crucial role in facilitating a discursive shift from an *external*-technical conceptual paradigm to an *internal*-autonomous understanding of purposiveness, but of course, as physicians and naturalists engaged in empirical research, they were unable to provide (or uninterested in providing) a philosophical account of this shift. This account was instead provided by Schelling and Hegel, who philosophically interrogated the self-organizing features of living systems. When asked for his opinion on such developments, Kielmeyer reported to Cuvier that 'through the newer philosophical systems, we have also become more accustomed than hitherto to consider nature in general and as a whole as a living organism in all its operations, and single organizations as individual representatives of nature at large'(p. 75). The point I have tried to make here is that this shift can be profitably reconsidered today through categories of twentieth-century cybernetics, such as homeostasis, goal-directedness and control.

(VI) Conclusions

I opened this essay by asking in what sense Kielmeyer can be considered to be the father of Romantic *Naturphilosophie*, as his former student Cuvier suggested in a lecture course of 1832. I submitted that this question can have two possible answers: a *developmental* and a *physiological* account of Kielmeyer's legacy. The former has generally received more attention via the works of Stephen Jay Gould and Robert J. Richards, whereas the latter has only recently begun to be addressed in the English-speaking literature. The former tradition attributes Kielmeyer's importance to the final statements of his 1793 Speech concerning the relation between individual development and the functional organization of the animal kingdom as an 'archaeology' (in Foucault's sense) of Haeckel's biogenetic law (according to which 'ontogeny recapitulates phylogeny') and ultimately of Evo-Devo. The latter focuses rather on Kielmeyer's general approach to the distribution of 'organic forces' in the organic world.

My key point has been that this latter approach can be re-interpreted today as a cybernetic model of nature conceived as a 'universal organism' that organizes and regulates itself through 'teleological mechanisms' not dissimilar in nature from those talked about by the participant of the Macy conferences. These teleological mechanisms, in turn, contribute to generate what German Idealists referred to as *Geist*, an 'ecology of mind' grounded on the notions of organization, homeostasis, information and control. I argued that this interpretation allows us to understand how the German Idealists grappled with the problem of teleology left open by Kant's 'antinomy of teleological judgment'. In particular, their adherence to a *radical* interpretation of the antinomy, which urges that, if philosophy intends to tackle Kantian 'intrinsic teleology' as a constitutive self-organizing process in nature, it rather needs to move *beyond* Kant's deflationism.

On the background of the German physiological tradition in the second half of the eighteenth century, I have thus tried to provide a contemporary reading of

Kielmeyer's laws concerning the inverse relations among 'organic forces' in nature. My reading has stressed how Kielmeyer's importance lies in the development of a cybernetic-systemic approach, which was enthusiastically taken up in Schelling's *World-Soul* and *First Outline*. I believe that the fundamental sense of this attempt goes largely in the direction of a 'naturalization of subjectivity'. This naturalization is not to be understood in reductionist terms, but rather as the 'vital grounding' of mind as coextensive with the intrinsic goal-directedness of biological systems. In this perspective, we could say that Blumenbach's *Bildungstrieb*, with its purposive tendency towards organization, ultimately provides the hallmark for the purposive features we attribute to conscious action.

In this sense, if the quest of German Idealism, as Beiser famously emphasized, should ultimately be understood as a 'struggle against subjectivism', then Grant is perfectly right in identifying Schelling as the early champion of a 'geology of morals'[60] to which Kielmeyer contributed by setting out its naturalistic foundations. Based on the above, I interpret this 'geology of morals' as a label for the general quest to overcome the Kantian dualism between 'concepts of nature' and 'concepts of freedom' and, with it, every abstract understanding of the self as detached from the overall ecological forces that contribute to generate it in the first place. German Idealists called those ecological forces 'the absolute' – and this, in my view, is just another name for 'the *systemic* nature of the mind'.[61]

Notes

1. T. Bach, 'Kielmeyer als "Vater der Naturphilosophie?" Anmerkungen zu seiner Rezeption im deutschen Idealismus', in K. T. Kanz (ed.), *Philosophie des Organischen in der Goethezeit: Studien zu Werk und Wirkung des Naturforschers Carl Friedrich Kielmeyer (1765-1844)* (Stuttgart: Franz Steiner, 1994), 234.
2. I would like to thank Luca Corti and Oriane Petteni for their thoughtful comments on the early draft of this paper.
3. S. J. Gould, *Ontogeny and Phylogeny* (Harvard: Harvard University Press, 1977).
4. W. Coleman, 'Limits of the Recapitulation Theory: Carl Friedrich Kielmeyer's Critique of the Presumed Parallelism of Earth History, Ontogeny, and the Present Order of Organisms', in *Isis* 64.3 (1973).
5. L. Oken, *Abriß des Systems der Biologie* (Göttingen, 1805), 1.
6. T. Bach, *Biologie und Philosophie bei C. F. Kielmeyer und F. W. J. Schelling* (Stuttgart: Frommann-Holzboog, 2001).
7. See Gould, *Ontogeny and Phylogeny*.
8. S. J. Gould and R. Lewontin, 'The Spandrels of San Marco and the Panglossian Paradigm: A Critique of the Adaptationist Programme', in *Proceedings of the Royal Society B* 205.1161 (1971).
9. See G. Fusco and A. Minelli, *Evolving Pathways: Key Themes in Evolutionary Developmental Biology* (Cambridge: Cambridge University Press, 2008); P. Huneman, 'Assessing the Prospects for a Return of Organisms in Evolutionary Biology', in *History and Philosophy of the Life Sciences*, 32.2-3 (2010); M. Pigliucci and G. Müller, *Evolution: The Extended Synthesis* (Cambridge, MA: MIT Press, 2010); A. Love,

Conceptual Change in Biology: Scientific and Philosophical Perspectives on Evolution and Development (Dordrecht: Springer, 2015).
10 See R. J. Richards, *The Romantic Conception of Life: Science and Philosophy in the Age of Goethe* (Chicago: University of Chicago Press, 2002); R. Amundson, *The Changing Role of the Embryo in Evolutionary Thought: Roots of Evo-Devo* (Cambridge: Cambridge University Press, 2005).
11 C. F. Kielmeyer, *Über die Verhältnisse der organischen Kräfte untereinander, in der Reihe der verschiedenen Organisationen, die Gesetze und Folgen diese Verhältnisse*, ed. K. T. Kanz (Marburg: Basiliken-Presse, 1993).
12 K. T. Kanz (ed.), *Philosophy des Organischen in der Goethezeit: Studien zu Werk und Wirkung des Naturforschers Carl Friedrich Kielmeyer* (Stuttgart: Steiner, 1994).
13 A. Gambarotto, *Vital Forces, Teleology, and Organization: Philosophy of Nature and the Rise of Biology in Germany* (Dordrecht: Springer, 2018); J. H. Zammito, *The Gestation of German Biology* (Chicago: University of Chicago Press, 2017); J. Steigerwald, *Experimenting at the Boundaries of Life: Vitality in Germany around 1800* (Pittsburgh: University of Pittsburgh Press, 2019).
14 G. Bateson, *Steps to an Ecology of Mind* (Chicago: University of Chicago Press, 1972), 434.
15 Ibid., 435.
16 Ibid., 482.
17 Ibid., 490.
18 Ibid., xxx.
19 Ibid., xxxii.
20 Ibid., 462.
21 I. H. Grant, *Philosophies of Nature after Schelling* (London: Continuum, 2006), 3–4.
22 An idea shared by organizational accounts of functions and teleology (M. Mossio et al., 'An Organizational Account of Biological Functions', in *British Journal for the Philosophy of Science*, 60.4 [2009]; M. Mossio and L. Bich. 'What Makes Biological Organization Teleological?' in *Synthese* 194.4 [2017]), enactivist approaches to cognition (S. Gallagher, *Enactivist Interventions: Rethinking the Mind* [Oxford: Oxford University Press, 2017]) and advocates of an Extended Evolutionary Synthesis (G. B. Müller, 'Why an extended evolutionary synthesis is necessary' in *Interface Focus* 7.5 [2017]), no less than by Schelling-inspired philosophers.
23 Grant, *Philosophies of Nature*, 10.
24 Bateson, *Steps*, 143.
25 Grant, *Philosophies of Nature*, 145, quoting Schelling's *Ideas for a Philosophy of Nature*.
26 I. Kant, *Critique of the Power of Judgment*, trans. P. Guyer (Cambridge: Cambridge University Press, 2000), 246.
27 Ibid., 271.
28 Ibid., 258–9.
29 T. Lenoir, 'Teleology without Regrets. The Transformation of Physiology in Germany: 1790–1847', in *Studies in History and Philosophy of Science, Part A*, 12.4 (1981).
30 K. L. Caneva, 'Teleology with regrets', in *Annals of Science* 47 (1990).
31 A. Breitenbach, 'Teleology in Biology: A Kantian Approach', in *Kant Yearbook*, 1.1 (2009), 31.
32 Kant, *Critique of Judgment*, 245.
33 P. Huneman, 'Kant's Concept of Organism Revisited: A Framework for a Possible Synthesis between Developmentalism and Adaptationism?', in *The Monist*, 100.3 (2017).

34 J. H. Zammito, 'Teleology then and now: The question of Kant's relevance for contemporary controversies over functions in biology', in *Studies in History and Philosophy of Biological and Biomedical Sciences*, 37.4 (2006), 748.
35 A. Weber and F. Varela, 'Life after Kant: Natural purposes and the autopoietic foundations of biological individuality', in *Phenomenology and the Cognitive Sciences* 1 (2002), 98.
36 H. Van den Berg, 'The Wolffian Roots of Kant's Teleology', in *Stud Hist Biol Biomed Sci* 44 (2013).
37 G. W. F. Hegel, *The Science of Logic*, trans. A. V. Miller (Amherst: Humanity Books, 1969), 654.
38 See F. Chiereghin, 'Finalità e idea della vita: la ricezione hegeliana della teleologia di Kant', in *Verifiche*, 19.1 (1990).
39 A. Rosenblueth, N. Wiener and J. Bigelow. 'Behavior, Purpose and Teleology', in *Philosophy of Science, 10* (1943).
40 See N. Wiener, *Cybernetics: Or Control and Communication in the Animal and the Machine* (Cambridge, MA: MIT Press, 1948).
41 G. Bowker, 'How to Be Universal: Some Cybernetic Strategies, 1943–70', in *Social Studies of Science* 23.1 (1993), 115.
42 See F. Heylighen and C. Joslyn. 'Cybernetics and Second-Order Cybernetics', in, R.A. Meyers (ed.), *Encyclopedia of Physical Science & Technology* (New York: Academic Press, 2001). A pivotal role in the genesis of the cybernetic approach was played by the so-called 'Macy conferences' held between 1946 and 1953, whose main goal was to turn the notion of 'teleological mechanisms' developed in the 1943 paper into a universal approach to be applied in the most different domains of natural, human and social sciences. The participants to the conferences included (among many others) social scientist Gregory Bateson, engineer Julian Bigelow, neuropsychiatrist Warren McCulloch, anthropologist Margaret Mead, philosopher Filmer Northrop, along with mathematicians John von Neumann, Walter Pitts and Norbert Wiener (see J. Heims, *The Cybernetic Group* [Cambridge MA: MIT Press, 1993]). These interdisciplinary scholars were convinced that the language of feedback and control mechanisms would ultimately contribute to break down 'the false dichotomies between mind and matter, human and non-human – dichotomies that the new information-based language would show never to have been true' (Bowker, 'How to Be Universal', 117). Within the scope of this mission, machines were used as experimental tools to test hypotheses on complex systemic phenomena, not as the initial template on which living organisms should be understood.
43 Heims, *Cybernetic Group*, 264.
44 Ibid., 236.
45 Ibid., 241.
46 Ibid., 246.
47 F. W. J. Schelling, *Von der Weltseele. Eine Hypothese der höhern Physik zur Erklärung des allgemeinen Organismus* (Stuttgart: Frommann-Holzboog, 2000), p. 252.
48 Ibid., 253.
49 N. Jardine, *The Scenes of Inquiry: On the Reality of Questions in the Science* (Oxford: Clarendon Press, 1991).
50 See R. Rey, *Naissance et développement du vitalisme en France de la deuxième moitié du 18ᵉ siècle à la fin du Premier Empire* (Oxford: Voltaire Foundation, 2000).
51 See T. Lenoir 'The Göttingen School and the Development of Transcendental Naturphilosophie in the Romantic Era', in *Studies in the History of Biology* 5 (1981).

52 S. A. Roe, *Matter, Life, and Generation Eighteenth-Century Embryology and the Haller-Wolff Debate* (Cambridge: Cambridge University Press, 1981).
53 J. F. Blumenbach, *Über den Bildugstrieb* (Göttingen: Dieterich, 1791), 12–13.
54 Ibid., 14.
55 The total count of vital forces in the *Institutiones physiologicae* is actually five: 1) the *formative drive* (*nisus formativus*), responsible for developmental organization and active throughout the life of the organized body; (2) *contractility* (*vis cellularis*), responsible for the contraction of mucosa; (3) *irritability* (*vis muscularis*), responsible for the contraction of muscles; (4) *vitae propriae,* responsible for the contraction of specific organs, such as the iris or the fallopian tubes; and (5) *sensibility* (*vis nervea*), which facilitates the perceptive functions. From a conceptual point of view, however, 'contractility' and 'vitae propriae' are just special cases of irritability. In a similar way, in the 1793 speech, Kielmeyer adds 'propulsive force' and 'secretive force,' which play absolutely no role in the speech itself. We can thus consider these as variations on the core organic forces constituted by sensibility, irritability and reproduction.
56 S., Schmitt, *Les forces vitales et leur distribution dans la nature : Un essai de ' systématique physiologique'*, text for C. F. Kielmeyer (1765–1844), H. F. Link (1767–1851) et L. Oken (1779–1851) (Turnhout: Brepols, 2006) ; S. Schmitt, 'Succession of Functions and Classification in post-Kantian Naturphilosophie around 1800', in P. Huneman (ed.), *Understanding Purpose. Kant and the Philosophy of Biology* (Rochester: University of Rochester Press, 2007), 123–36.
57 R. J. Richards, 'Kant and Blumenbach on the *Bildungstrieb*: A Historical Misunderstanding', in *Stud. Hist. Phil. Biol. Biomed. Sci.* 31.1 (2000).
58 F. W. J. Schelling, *Erster Entwurf eines Systems der Naturphilosophie* (Stuttgart: Frommann-Holzboog, 2000), 116.
59 Ibid., 211.
60 Grant, *Philosophies of Nature*, 8.
61 Bateson, *Steps*, 145.

Chapter 12

REPRODUCTION, PRODUCTION AND THE EARTH: THE PLACE OF SEX IN KIELMEYER'S 'ECONOMY OF THE ORGANIC WORLD'

Susanne Lettow

As Kielmeyer explains in his *Ideas for a Universal History and Theory of the Developmental Appearances of Organisations* (1793/94), his overall project is to deduce the laws and determine the relations of forces of the 'life and economy of the organic world' (*GS*, 123). Accordingly, the main epistemic object that he deals with in this text, as well as in his 1793 Speech, is not the forces that constitute living beings but the 'relations of their distribution in different organisations' (*GS*, 122), as well as the changes of these relations with regard to the various temporal developments within the organic world. In order to explain the regularity of the changes of relations and the specific proportions of forces that characterize living beings, Kielmeyer, in his 1793 Speech, formulates a 'universal law' (p. 41), according to which the forces of irritability, sensibility, reproduction, secretion and propulsion compensate each other so that different organisms are characterized through different proportions or modes of equilibrium. He introduces this law with reference to his distinction of the five forces and three specific laws that govern the reproductive force, which he considers to be 'by far the most universal' force, and the one 'lavished in the greatest measure on the organisations' (p. 38). Whereas, in *On the Formative Drive* (1781), Johann Friedrich Blumenbach, had claimed that nutrition, regeneration and reproduction are just different manifestations of that fundamental drive,[1] Kielmeyer emphasizes the difference between the five forces and the peculiar status of the reproductive force.[2]

With his reflections on the nature of this force, Kielmeyer certainly contributed to the process by which, at the turn of the nineteenth century, 'the reproduction of organic beings, in contrast to their mere production, became recognizable as a domain governed by laws of its own'.[3] Moreover, his 1793 Speech has been designated 'the first system programme of biology'[4], since Kielmeyer here envisions the possibility of conceiving of living beings and their interrelations beyond the epistemic restrictions that Kant had formulated in his *Critique of Judgement*.[5] Indeed, for Kielmeyer, the life of the individual is integrated within the 'system of effects that one may call the life of the species', and this within 'the system of effects that constitutes the life of the great machine of the organic world' (p. 30), so that organisms, species and organic nature in its entirety form different but interrelated levels of a dynamic continuum.[6] Thereby, as Peter Hanns Reill notes, Kielmeyer does not assume 'a simple unified developmental pattern ... [that accounts] for species change and permanence', but his dynamic understanding of organic nature is based on the idea of multiple temporalities.[7] Accordingly, the reproductive force that is the main object of epistemic scrutiny manifests itself in multiple forms: it can appear here 'in one form eternally, there, in the form of a changeable fairy. Here it persists for centuries unworn, there its effects are contained in the passing of a moment [*Zeitfluxion*]' (p. 38).

Although reproduction obviously plays a central role in Kielmeyer's theory of organic nature, sex is only a marginal issue. Nevertheless, Kielmeyer refers to the issue on several occasions in his writings – and I will explore these references in this chapter. In particular, I want to argue that an analysis of the way in which Kielmeyer deals with the question of sex and sex differences not only adds a specific thematic element to the study of his thinking, but also sheds light on the debate about Kielmeyer's status within the history of the life sciences and German *Naturphilosophie*. In what follows, I distinguish between three different aspects of his thought. First, I reconstruct Kielmeyer's understanding of the gradual development of sex differences in different species. Secondly, I show that for Kielmeyer the developmental process itself, i.e. the process that moves from a state of indifference to complex differentiation, is gendered. Third, I will discuss Kielmeyer's remarks on the earth as a female procreative force. Certainly, Kielmeyer made no effort to systematize his reflections on sex or sex differences or to formulate a coherent theoretical position. However, his reflections converged with those of other authors – in particular, Goethe – and were adopted and re-articulated in various ways by authors such as Schelling, Oken and Hegel. Thus, despite all epistemic and political-ethical differences, Kielmeyer's ideas circulated within the discourse of *Naturphilosophie*.

The Gradual Development of Sex: Differentiation and Polarization

In his 1793 Speech, Kielmeyer states that the reproductive force manifests itself in a 'richness of shapes' (p. 38) while it is governed by 'a few very simple laws' (p. 38). According to the first law, the number of new individuals 'increases as the size of the producing individuals, or more universally, the individuals produced as they appear after birth decreases' (p. 39). According to the second law, an increase in the number of new individuals correlates to a decrease in the complexity of their physical structure and to a shorter period necessary for their formation. The third law finally states that

> the less the reproductive force is expressed in many new individuals, the more it is expressed either through metamorphoses that the body withstands, or

through unusual artificial reproduction, or both together, or by indeterminate growth, or by greater deviation in the newly begotten formations.

pp. 40–1

What Kielmeyer has in mind, when he writes about 'deviations' are sex differences. So, in his view it is 'still noteworthy, that in exactly those mammals and birds where the least productivity prevails, the newly produced individuals are most differentiated in their genitalia' whereas, in contrast, 'in all other classes – that display metamorphosis, unrestricted growth, or great reproduction of parts as is manifest in many insects and worms and so on – both sexes become more similar or disappear altogether' (p. 40). Obviously, Kielmeyer conceives of sex differences as a gradual phenomenon that can even be completely absent, and this corresponds with his reflections on the multiple forms of reproduction, including asexual reproduction. For example, Kielmeyer discusses *generatio aequivoca* at length in his inaugural speech at the University of Tübingen in 1796 on *The Origin of Infusoria*. 'Generation without parents,' he states, happens 'when all other forces have ceased to manifest themselves and have confined themselves to the reproductive force as the only one remaining' (GS, 33). This is the case, Kielmeyer continues, in death, i.e. the corpse, because the reproductive force is the force 'that die[s] off the last in the series of organizations' and also in every individual organism (GS, 33).[8] It follows that sexual reproduction is thus a highly specific form of bringing forth new individuals and Kielmeyer certainly does not conceive of it as a constitutive feature of the organic world. Accordingly, he argues in his 1790 *Outline* that metamorphosis unfolds differently in different plants, so that only in some does it result in 'a distinct flower and detectable sexual organs' (GS, 16).[9] In a similar vein, he discusses the gradual development of sex differences in the 1793/4 *Ideas*. There he contends that during the 'period of becoming', i.e. during the formation of the embryo, those organs that are formed 'in greater temporal distance' (GS, 154) from each other, such as the brain and the sexual organs, display fewer similarities than the extremities which are formed simultaneously. Kielmeyer thus stresses again, first, that physiological sex differences are the result of a developmental process, and, secondly, that embryonic development does not follow a single uniform trajectory. The sexual organs, he puts it, 'never acquire their fully developed form and very often not even their existence' (GS, 185) during the period of embryonic development.

Obviously, Kielmeyer assumes that asexuality characterizes the simplest organisms in which the reproductive force is abundant, and when sexual differentiation happens it does so in different forms. At first glance, this seems to stand in sharp contrast to Schelling's idea of nature as basically structured through a sexual dualism based on a bifurcation and sexualization of Blumenbach's concept of the formative drive. For Blumenbach, nutrition, regeneration and procreation were 'essentially just modifications of one and the same force'[10], and, although he referred to maternal and paternal reproductive fluids, Blumenbach did not conceptualize sex differences. Schelling, however claims that the formative drive 'splits in opposite directions, which on a lower level will appear as differentiation of the sexes'.[11] This means that, for Schelling, nature is inherently sexed, or as he puts it: 'an *a priori* regulative principle requires that sex difference[12] be taken as point of departure everywhere in organic nature'.[13] On this assumption, Schelling conceives of nature as an infinite process of reproduction in which the male and the female unite in order to produce a new entity that again is male or female and thus predisposed to unite with its complement. This might seem to be a radical departure from Kielmeyer's account, which starts not from an *a priori* dualism of the sexes but emphasizes the plurality of the ways in which the force of reproduction manifests itself. The contrast between Schelling's and Kielmeyer's accounts of sex differences, however, appears less sharp if one takes into account Schelling's references to Kielmeyer's understanding of gradual development. According to Kielmeyer, the organic world is structured through an increase in complexity, which at the same time is a process of differentiation that most prominently manifests itself in differences between the sexes. In the *First Outline of a System of the Philosophy of Nature*, Schelling draws on this idea by claiming that 'the separation into different sexes happens for different organisms at different stages of formation'.[14] By paraphrasing Goethe's essay on metamorphosis, Schelling argues that in plants, the 'development of the sexes is merely the highest zenith of the process of formation'.[15] In addition, he contends, with regard to organisms such as insects, that 'in the first stage of their development no sexual difference shows itself'[16], but that this difference only emerges in and through the process of metamorphosis. Thus, even if sex difference, for Schelling, is a constitutive feature of all organisms, it acquires the form of polar opposition only at the highest stage of the development of organic nature. At this stage, he argues, the 'opposing natural activities'[17] that shape the organism 'arrive at the maximum point of mutual independence'.[18]

Indeed, Goethe's *Essay on the Metamorphosis of Plants* (1790) is an important intertext for understanding both Kielmeyer's and Schelling's accounts of graduation and sex difference. In this text, Goethe formulates the idea of a gradual dynamic development of sex differences, arguing that metamorphosis 'ascends' to the 'summit of Nature, the reproduction through two sexes'.[19] Like Kielmeyer, Goethe introduces the idea of a correlation between gradual development and differentiation, or, as Jocelyn Holland argues, sequential and simultaneous development. 'The appearance of the sexes,' she states, 'disturbs the linearity of plant development and makes a linguistic demand on the observer to account for things which occur both sequentially and simultaneously'.[20] Although for Goethe the process of the metamorphosis of plants is geared towards the emergence of a binary sexual difference, he stresses 'anastomosis', i.e. the mutual connection of the sexes in copulation and the 'kinship' of the male and the female organs.[21] Like Kielmeyer, Goethe thus does not translate difference into polarity and hierarchy in this early text. As Stefani Engelstein puts it with regard to the *Metamorphosis of Plants*, Goethe's

> attitude towards sexuality is … complex, involving an understanding of sexual division as a higher order function tending towards the service of male *Bildung*,

and yet entailing enough flexibility for a high level of correspondence and non-sexed phenomena.[22]

However, even if Kielmeyer and Goethe, in their writings from the 1790s, do not emphasize the idea of a sexual dualism, the idea that a binary sex difference emerges at the highest stage of organic development and thus represents the most complex organic structure was re-interpreted by subsequent authors in terms of sexual polarity and hierarchical complementarity.[23]

These were not mis-readings, but interpretations that could find points of reference in the respective writings. According to Bersier, Kielmeyer not only introduced the idea of 'dynamic graduation', but also construed 'his two principles of sensitivity and reproduction ... as polar opposites'.[24] This statement might be overblown, since Kielmeyer conceives of five interrelated forces and not a dual structure; however, Bersier captures a hierarchical logic that nevertheless is at play in Kielmeyer's text. As Holland explains with regard to Novalis: the distinction between sensitivity and reproduction

> was taken up as a way of describing the balance between physical and intellectual faculties and to claim man's specificity as the creature with the greatest sensitivity: whose lessened physical reproduction, when compared to other species, was compensated by the production of ideas.[25]

In addition, this distinction – which corresponds to the Aristotelian distinction between matter and form – was gendered, and transposed into an argument for the supposed intellectual superiority of men over women, who were deemed to be bound up with reproduction[26]

A Gendered Development: From Indifference to Difference

Already Jean-Baptiste Robinet had formulated the idea that 'nature has always proceeded from the least composite to the more composite. The most complex organization that we know, and that which produces most phenomena, is that of man'.[27] In addition, Herder had developed the idea that, in the more complex living beings, the force of reproduction is weaker whereas sensitivity is stronger, and, in the human, develops into reason, while at the same time in the less complex organisms, generation and regeneration prevail. Kielmeyer builds on these authors and lingering notions of gender, which become recognizable at the margins of his texts. The second aspect of Kielmeyer's thought that is relevant both for determining the place of 'sex' in his account of organic nature and for scrutinizing the adaptations and transformations of Kielmeyer's thinking by naturalists and philosophers of the period consists in the gendering of the basic concepts and ideas through which he articulates his account. Indeed, in Kielmeyer's writings, gradual development or the process of differentiation that produces an increase in complexity, itself is gendered. So, in the *Outline*, Kielmeyer contends that the female sexual organs become more important and persistent 'downwards in the series of the animals' whereas the male organs 'become more and more similar to the female ones, and finally cease to exist in their particularity within a separate individual' (*GS*, 20). The most simple, undifferentiated state of the organic thus appears to be female. As a consequence, indifference and differentiation, simple and complex, female and male, are constituted as homologous distinctions that are meant to reveal the basic structure of the 'economy of the organic world'.

Although Kielmeyer neither reflects nor repeats these homologies extensively, naturalists and philosophers such as Lorenz Oken and Hegel adapted and transformed them in particular ways. I focus on their accounts here because they reveal, first, that the relation between the distinction of male/female and the distinctions difference/indifference and complex/less complex is completely arbitrary, and, secondly, that, regardless of the discursive distribution of masculinity and femininity, these homologies articulate hierarchical views that fostered patriarchal gender relations. In *Die Zeugung* (1805), Oken attempts to present a new theory of generation based on his notion of infusoria, which he considers to be the 'primordial substances (*Urstoffe*)' of which all organisms are made.[28] In so doing, he distinguishes between three stages of organic development, i.e. the stages of the polyp, the plant and the animal. At the most basic organic stage, that of the infusoria, sex differences are absent because, he contends, reproduction happens without the interference of a female contribution. The infusoria, Oken writes, are

> an infinite growth without any leap, and fragment themselves by inner force, without needing the help of a complementary individual, not into such an infusorium with which they would enter into a relation of any duplicity but into an individual that equates the first and all other ones absolutely in scope and function without any sex difference.[29]

Oken considers this to be the 'male' stage of the polyp.[30] At the second stage, Oken contends, bifurcation and thus differentiation first take place. He conceives of this stage of the plant as 'female' because, with bifurcation, independence is lost, and, for Oken, the female as such is per se dependent. The plant, he wrote, exists 'only through the infusorium, the female lives only through the male such as the plant though the polyp'.[31] At the highest stage of animal formation, and finally the formation of the human, the male principle of autonomous reproduction, again, holds sway over the female. In Oken's view, this again is a 'male' stage because the female principle is subjugated. He thus finally and infamously concludes that the 'total human species is nothing but a continually propagating male'.[32] This threefold process that proceeds from male indifference to female difference to male autonomy that integrates both indifference and difference is overtly gendered and articulates a masculinist worldview that clearly differs from that of Kielmeyer.

In addition, it also differs from Hegel's gendered articulation of the distinction between indifference and difference. In his philosophy of nature, Hegel mocks Oken's notion of the infusoria as 'an amorphous jelly, an active slime which is reflected into itself and in which sensibility and irritability are not yet separated'.[33] However for Hegel, too, the lowest stages of the organic world are undifferentiated. Hegel conceives of plants as sexless living beings that do not reproduce but just 'produce' new organisms. This is the case because, as he argues,

> the plant does not attain to a relationship between individuals as such but only to a difference, whose sides are not at the same time in themselves whole individuals, do not determine the whole individuality; therefore the difference, too, does not go beyond a beginning and an adumbration of the genus-process.[34]

In Hegel's view, sex difference in the proper sense of the word only emerges at the stage of animal nature, which means that only at this stage is organic nature characterized through a 'veritable separation of the sides of the opposition in the sex-relation'.[35] While Hegel thus also conceives of a development from a stage of sex indifference to one of sex difference, his attribution of masculinity and femininity to these stages is antithetical to that of Oken. Due to the central role that bifurcation and differentiation play in Hegel's overall philosophical narration, differentiation is highly valued. Without differentiation negation is impossible and so is the dialectical process that Hegel conceived of. Accordingly, Hegel associates indifference with female passivity, while he conceives of differentiation as an active 'male' process. The sexes, in Hegel's view, share an 'original identity' (ibid.) which, however, is expressed differently. While for the woman the essential is that which is 'indifferent',[36] for the man it is 'the moment of duality, the opposition'.[37] Hegel's description of the male and female sexual organs in terms of the opposition between passivity and activity reinforced this. Thus, he claims that

> just as in the male, the uterus is reduced to a mere gland, so, on the other hand, the male testicle remains enclosed in the ovary in the female, does not emerge into opposition, does not develop on its own account into active brain, and the clitoris is inactive feeling in general.[38]

In both cases, for Oken and Hegel the distinction between indifference and difference, and thus the developmental process from one stage to the other that constitutes the organic world, is obviously gendered. The clear gender hierarchies that Hegel translates into his political philosophy[39] can be found neither in Kielmeyer nor in Schelling who, more similarly to Goethe, stresses the unification of the sexes and the provisional disappearance of sex differences in the process of reproduction. 'Nature,' as Schelling puts it in the *First Outline*, 'hates sex' because 'it is always striving only for a return into the identity of the genus, which however, is enchained to the (never to be cancelled) duplicity of the sexes, as to an inevitable condition'.[40] The second aspect of Kielmeyer's understanding of sex difference, i.e. the idea that the process of gradual differentiation is gendered, was as well adopted, emphasized and thereby transformed into an explicitly hierarchical understanding of gender differences.

The Productivity of the Earth: Symbolic Femininity

The third aspect of Kielmeyer's articulation of sex that I want to explore is the evocation of symbolic femininity in his references to the earth. In Kielmeyer's thought, the productivity of the earth plays an important role, in particular in the context of his idea of recapitulation. In the 1793 Speech, Kielmeyer stresses that 'the distribution of the forces in the series of organizations follows the same order as the distribution in the different developmental states of the same individual' and that 'the force by which generation occurs in the latter, namely the reproductive force, is consistent in its laws with the force by which the series of different organizations of the earth are called into existence' (p. 43). Kielmeyer shifts the perspective from a homology regarding the distribution of forces to an accordance of the laws of the reproductive force with the laws of the force that accounts for the production of life on earth, and then concludes that one can assume a single 'material cause' (p. 43) for the explanation of the different developmental stages and for the first production of organisms on earth. As Iain Hamilton Grant argues, it is crucial to keep in mind that Kielmeyer does not conceive a parallelism of bodies or somatic features but a parallelism of forces and laws of forces.[41] Indeed, Kielmeyer's argument moves from the distribution of forces to the relation of the laws of forces to the assumption of a single productive force that he understands as being the 'material cause' of the development of individuals, the development of the 'series of species' and 'the first production of organizations on our Earth' (p. 43). He introduces this parallelization of forces and laws of forces most prominently in the 1793 Speech, but refers to the idea of the generative capacity of the earth itself in other writings, too. In the 1793/94 *Ideas*, for example, he emphasizes that 'the procreating, more complex organism relates to the procreated, dissimilar and more simple one almost in the same way as our generating earth may have once related to the generated series of organizations' (*GS*, 131). And, in his letter to Windischmann from 25 November 1804, he then claims again that he holds

> the force by which the series of organic bodies was once brought forth on our earth, to be one and the same as regards its essence and laws, as that force by which even now the series of developmental stages is produced in every organic individual, which is identical to that of the series of organizations.
>
> p. 65

This force is clearly addressed here as a female force because Kielmeyer links it to the notion of the earth as mother. The results of the productivity of the earth are, as he put it, 'the children of the earth' (p. 65) or 'children of the life-giving [gebährende] earth' (p. 67).

With regard to the meaning of 'sex' in Kielmeyer's writings, this means that the ideas of the interrelation between graduation and differentiation, and of the gendered relation between indifference and difference, are supplemented by the idea of an all-encompassing feminine force. He thereby evokes the ancient mythological notion of the earth as female, most prominently represented as the goddess Gaia. Kielmeyer's references to the idea of the female earth were not unique or exceptional, however. In the context of the 'temporalization' of natural history in the second half of the eighteenth century,[42] authors such as Benoît de Maillet and Julien Offray de La Mettrie also ventured far into cosmological speculation, amalgamating science and mythology in new ways. In his *Système de L'Epicure* of 1750, La Mettrie for example mused that in its earliest history 'the earth must have served as a uterus to man; it must have opened its womb to the human germs, already prepared, so that given certain laws, this superb animal could hatch out'.[43] Even Kant admitted that he sometimes felt compelled to similar speculations, which he considered to be a 'daring venture on the part of reason' but not *a priori* 'absurd'.[44] In §80 of the *Critique of Judgement*, he muses about the possibility of 'an actual kinship' among animal species, reaching 'from that in which the principle of ends seems best authenticated, namely from man, back to the polyp, and from this back even to mosses and lichens, and finally to the lowest perceivable stage of nature ... to crude matter'.[45] Such a 'great family of living things', as Kant put it, could be conceived as originating from 'a universal mother [*gemeinschaftliche Urmutter*]' or from the 'womb of mother earth as it first emerged ... from its chaotic state'.[46] Like Kant, Kielmeyer distances himself from his inclination to cosmological speculation when, with respect to a history of emergence of life on earth, he writes to Windischmann that he doubts whether

> the execution of such a history is possible ... primarily because we *know nothing* about the *development of our earth and its magnetism* and therefore regarding a law of these developments, and just as little concerning a *law to which the gradation of existing* organizations conforms.
>
> p. 66

Nevertheless, the female Earth plays a relevant role in his thought because Kielmeyer, through this idea, articulates the overall cohesion of the economy of the organic world, its basic 'material cause'.

Schelling, as I have argued, adopts Kielmeyer's ideas about the gradual development of sex differences, while he emphasizes the duality of the sexes and re-articulates it in the context of his philosophy of nature as an *a priori* principle of the organic world. A similar move can be observed with regard to the idea of a feminine-motherly basic structure. Where Kielmeyer simply alludes to this idea and – like Kant – refrains from rendering it into an explicit theoretical claim, Schelling transposes this idea into philosophy. This means that, through amalgamating Kielmeyer's 'material cause' with Plato's notion of *chora*, Schelling transforms the idea into a central element of his philosophical position. In the *Philosophical Investigations into the Essence of Human Freedom*, Schelling explicitly refers to the *Timaeus* in which Plato introduces the notion as a 'third genus' that comprises 'the being which is indivisible and remains always the same and the being which is transient and divisible in bodies'.[47] For Plato, the *chora*, which he calls the 'wet nurse of becoming'[48] is 'the formless, primal material space that he ... explicitly describes in feminine terms'.[49] Schelling re-articulates the *chora* as the 'non-ground [*Ungrund*]' that precedes all opposites, which 'cannot be distinguishable in it nor ... present in any way'.[50] In *On the World Soul* (1798), he addresses the *chora*, paraphrasing Plato, as the 'principle of life' which is 'although receptable for all forms, itself originally formless (αμοςΦον) and nowhere representable as positive matter'.[51] Whereas Kielmeyer, in his references to the productivity of the earth, evokes mythological undertones, Schelling introduces a philosophical principle that builds on ancient Greek philosophy and respective imaginations of the feminine, which he at the same time masculinizes.[52]

To sum up, Kielmeyer – apart from the remarks that I have mentioned – does not elaborate on the idea of a primordial feminine ground or matrix from which all differentiation springs, sexual or not. Like his remarks on the other aspects of sex and sex difference that I have discussed, his explicit references to this idea are marginal and often merely allusions. However, what makes Kielmeyer's articulation of sex interesting is that he integrates three different aspects of the meaning that together form a semantic structure that became a decisive matrix for modern notions of sex. According to this structure, 'sex' first refers to certain features and differences of organisms that appear in various shapes throughout the organic world. Secondly, the notion of sex is inextricably linked to a gendered hierarchy within the organic world that positions the female at the lower, the male at the higher stages. Thirdly, the notion of 'sex' is part of a mythological narrative according to which the order of the organic world in its totality derives from a feminine origin. It is thus the feminine itself that accounts for both the different forms that sex differences acquire within organic nature, as well as for the hierarchical structure that devalues everything that is female.

Conclusion

Recently, Zammito has drawn on Schelling's statement in *On the World Soul* (1798) that Kielmeyer introduces 'an entirely new epoch of natural history', arguing that this epoch 'saw the dawn of what would be called "biology" by several key figures in the very years Schelling was writing'.[53] In a similar vein, Iain Hamilton Grant – in his seminal book on *Philosophies of Nature after Schelling* – has highlighted Kielmeyer's post-Kantian turn. Kielmeyer, he argues, formulates an understanding of nature as a reality 'necessarily extending beyond any and all "possible experience"'.[54] Both readings stand in clear contrast to Reill's classification of Kielmeyer as 'an authentic representative of Enlightened Vitalism'[55], and thus as an opponent of *Naturphilosophie*. Reill refers to Kielmeyer's positive references to Kant's transcendentalism and his harsh critique of *Naturphilosophie* at the end of his letter to Cuvier from 1807. In addition, he argues that Kielmeyer develops a nuanced and complex understanding of sex differences that is distinct from later naturephilosophical polarizations. Like other authors in the late Enlightenment, Kielmeyer – Reill stresses – does not pay much epistemic attention to sex differences and refrained from dualistic thinking.[56] Indeed, there is much evidence that the emphasis on both polarity and hierarchy gained prominence only in the early nineteenth century in *Naturphilosophie* and the wider post-revolutionary discourse. Authors such as Karin Hausen, Londa Schiebinger, Thomas Laqueur or Claudia Honegger[57] have all shown that political and cultural claims to gender equality were highly contested in the period, and that references to 'nature', in particular within anthropology and biology, were part of new strategies for the justification and re-establishment of gender hierarchies.

So how to position Kielmeyer within these historical narratives? Does Schelling's claim that Kielmeyer initiated a 'new epoch' also hold for understandings of sex and gender differences? First and foremost, a careful reading of his texts challenges any linear understanding of cultural and epistemic change. In his letter to Cuvier from December 1807 that ends with the well-known statement that *Naturphilosophie* has 'done more harm than good to younger persons in Germany' (p. 76), Kielmeyer himself claims to have introduced a focus on polarity and dualism in natural history. In particular, he claims that 'already in the years 1790 to 1792, well before the philosophical systems after Kant were being discussed' (p. 75), he sought

> to prove, and was perhaps the first to credibly do so, that the vital operations in organic bodies in general and the developmental effects, especially as manifest in rigid and fluid parts, were to be ascribed to a force working dually towards its poles and acting as magnetism and electricity do.
>
> p. 75

In addition, he notes that this aspect of his thought had been taken up by subsequent, post-Kantian philosophers. This was, indeed, the case and led to a series of appropriations of Kielmeyer's thought that were basically transformations. Certainly, his texts contained reflections and ideas that, even where they were not systematized, proved fruitful for contemporary readers. Kielmeyer, however, neither introduced nor initiate a new epoch nor was he completely opposed to thoughts that gained prominence in *Naturphilosophie*. Instead of putting him on this or that side of the divide between Kantianism and post-Kantianism, Enlightenment and *Naturphilosophie*, it is more interesting to focus on the margins and ambivalences of his thought and on the heterogeneous, contested and contestable readings that it inspired.

Notes

1 Blumenbach himself distinguished between five forces in his *Institutes of Physiology* (1787). However, 'for thinkers of the 1790s it would not be so much these specific forces that proved most important. Rather, *Bildungstrieb* became the general term for the distinctiveness of living things ... in a manner that vastly exceeded the specificity of its creator's intentions.' J. H. Zammito, *The Gestation of German Biology. Physiology and Philosophy from Stahl to Schelling*, Chicago: Chicago University Press, 2017), 249.
2 Kielmeyer draws on Herder who, in his *Ideas for the Philosophy of the History of Humanity* (1784–91) also developed a theory of different forces, including the

reproductive force, and argued that the different forms of living beings express specific hierarchies of the forces. In particular, 'William Coleman and Timothy Lenoir ont signalé ce que Kielmeyer devait à Herder, le premier en insistant sur le rapport entre la géologie et la vie des espèces, le second plus directement à propos de la notion de récapitulation' R. Rey, 'La recapitulation chez les physiologists et naturalists allemands de la fin du XVIIIe et du début du XIXe siècle,' in P. Mengal (ed.), *Histoire du concept de recapitulation* (Paris: Masson, 1993), 47.

3   S. Müller-Wille and H.-J. Rheinberger, 'Heredity –The invention of an epistemic space', in Müller-Wille and Rheinberger (eds), *Heredity Produced: At the Crossroads of Biology, Politics, and Culture, 1500–1870*, Cambridge, MA: MIT Press, 2007), 6. A careful analysis of Kielmeyer's writings could provide important insights for a 'larger-scale map of practices, concepts and word usage' (N. Hopwood, 'The Keywords "Generation" and "Reproduction"', in Hopwood, R et al. (eds), *Reproduction from Antiquity to the Present* [Cambridge: Cambridge University Press], 304) that Nick Hopwood proposes in his evaluation of existing research on the shift 'from generation to reproduction' broadly indicated by Francois Jacob. Until now, Kielmeyer's status within the history of reproduction has not been systematically explored.

4   T. Bach, *Biologie und Philosophie bei C.F. Kielmeyer und F.W. J. Schelling* (Stuttgart-Bad Cannstatt: Frommann-Holzboog, 2001), 87.

5   As Thomas Bach has emphasized, Kielmeyer linked his physiological doctrine of forces to temporalized natural history so that, as a result, 'the realm of the living' emerged as 'as a domain of its own' (Bach, *Biologie und Philosophie*, 87).

6   Kielmeyer here, too, borrows both from Herder and from Robinet. 'En instaurant une continuité du minéral au vivant, Herder rejoint les idées de Jean-Baptiste Robinet qui, dans *De la Nature*, cherchait à retrouver dans la morphologie des minéraux et des plantes les ébauches de la forme humaine et certaines formes' (Rey, 'La recapitulation', 43).

7   P. H. Reill, *Vitalizing Nature in the Enlightenment* (Berkeley: University of California Press, 2005), 195. Zammito is certainly right when he states that 'Kielmeyer undertook what Arthur Lovejoy famously termed the "temporalization" of "the great chain of being"' (*Gestation*, 258). However, temporalization was not a one-dimensional process but included different and partly contradictory conceptualizations of temporality as well as – like in the case of Kielmeyer – the conceptualization of multiple temporalities. Therefore, the idea of temporalization needs to be unpacked. Cf. B. B. von Wülfingen et al. 'Temporalities of Reproduction: Practices and concepts from the eighteenth to the early twenty-first century. Introduction', in *History and Philosophy of the Life Sciences* 37.1 (2015): 1–16.

8   Kielmeyer assumes that water is the 'medium' that transmits the reproductive force that in the living body collides with the other forces and therefore manifests itself in 'the production of similar organisms', while it produces 'different organizations' in the dead body (*GS*, 33)

9   Kielmeyer here uses the concept of metamorphosis that Goethe adopted from Ovid in his *Essay on The Metamorphosis of Plants* (1790) for thinking through the developmental processes that shape living beings. Whether or not he had read Goethe's essay at this point is not clear. 'Undoubtedly,' as Gabrielle Bersier puts it, 'both Goethe's own discretion regarding his interest in Kielmeyer as well as Kielmeyer's own modesty have added to the mystery that shrouds their relationship' (G. Bersier, 'Visualising Carl Friedrich Kielmeyer's Organic Forces: Goethe's Morphology on the Threshold of Evolution', in *Monatshefte* 97.1 [2005], 19). While Kielmeyer's knowledge of Goethe's

writings remains obscure, Bersier and, based on her analysis, Zammito have explored Goethe's relation to Kielmeyer. Goethe read Kielmeyer's address of 1793 shortly after its publication and briefly visited Kielmeyer in 1797. 'Goethe returned to Kielmeyer's address in 1806 for a new reading, as he was preparing a comprehensive account of morphology with a view to publication. He returned to Kielmeyer's lecture notes in 1813, as once again he prepared to publish his morphological paradigm. Thus, Goethe fully recognized the epochal character of Kielmeyer's work and its dramatic convergence with his own' (Zammito, *Gestation*, 296). Gabrielle Bersier even 'suggests that Kielmeyer's address of 1793 occasioned "a turning point in Goethe's development"' ('Visualising Kielmeyer', 19).

10 J. F. Blumenbach, *Über den Bildungstrieb und das Zeugungsgeschäfte* (1781) (Stuttgart: Gustav Fischer, 1971), 19.

11 F. W. J. Schelling, *First Outline of a System of the Philosophy of Nature*, trans. Keith R. Peterson (Albany, NY: SUNY Press, 2004), 6.

12 Keith R. Peterson translates *Geschlechtsverschiedenheit* as 'sexual difference'. Here, and in other quotations, I prefer the term 'sex difference' in order to avoid a conflation with the notion of sexual difference as 'a specifically psychoanalytical concept, referring to subject positions within the symbolic order.' S. Sandford, *Plato and Sex* (London: Polity Press, 2010), 1.

13 Schelling, *First Outline*, 36.

14 Ibid. Indeed, Schelling concludes that 'the variety of organisms is finally reducible just to the variety of the stages at which they separate themselves into opposed sexes' (ibid., 42).

15 Ibid., 37.

16 Ibid.

17 Ibid., 40.

18 Ibid.

19 J. W. Goethe, *Scientific Studies* (Princeton: Princeton University Press, 1995), 87.

20 J. Holland, *German Romanticism and Science. The Procreative Politics of Goethe, Novalis and Ritter* (London: Routledge, 2009), 37.

21 Goethe, *Scientific Studies*, 84, 86.

22 S. Engelstein, 'Sexual Division and New Mythology: Goethe and Schelling', in S. Lettow and G. Rupik (eds), *Conceiving Reproduction in German Naturphilosophie*, forthcoming, 14.

23 Goethe himself developed the idea of polarity as a fundamental feature of organic nature in his later writings, and in collaboration with Schelling. 'Schelling had stressed magnetism and polarity in his 1797 work, *Ideen*. Goethe began magnetic experiments in the wake of reading that work. [...] In the summer of 1798 Kielmeyer's student Karl von Eschenmayer (1768–1852) published a text entitled *Versuch, die Grenze magnetischer Erscheinungen aus Sätzen der Naturmetphysik, mithin a priori zu entwickeln*. This would be one of the texts upon which Schelling would draw for his own *Naturphilosophie* and, with Eschenmayer and others, build it into a movement. The text came to the attention of Goethe almost immediately upon its publication. What attracted all these thinkers to magnetism was its essentially polar structure. Polarity had its literal seat, they believed, in magnetism' (Zammito, *Gestation*, 301).

24 Bersier, 'Visualising Kielmeyer', 21.

25 Holland, *German Romanticism and Science*, 93.

26 Joseph Görres, for example, contends that masculinity – which he understands as a cosmological principle – 'is the intellectual, the sensitive, the living which eternally

operates through producing and forming ...' while femininity represents 'external Nature who, in her womb contains the negative infinity of matter ... she who, continuously giving birth to that which she has conceived from the form, out of the absolute educes into reality.' J. Görres, 'Prinzipien einer neuen Begründung der Gesetze des Lebens durch Dualism und Polarität (Fortsetzung)' (1802) in *Naturwissenschaftliche, kunst- und naturphilosophische Schriften I (1800–1803)*, ed. Robert Stein (Köln: Im Gilde Verlag, 1932), 45.

27   Quoted in P. Bowler, 'Theories of Generation and the History of Life', in R. Stephenson and D. N. Wagner (eds), *The Secrets of Generation. Reproduction in the Long Eighteenth Century* (Toronto: Toronto University Press), 2015, 86. As Bowler explains, 'here what is, in effect, the panspermist version of the theory of preexisting germs becomes the basis for a model of historical progress predetermined by the forms prefigured in the germs that were created as the foundation for the universe itself. In another work (1768) Robinet attracted ridicule for suggesting that lower forms of life could be found anticipating the human form, mermen or aquatic humans being an example' (ibid., 87).

28   In these 'most simple living beings,' according to Oken, 'the chaos of Creation renews itself and disappears on a daily basis ... They emerge in every rot of organic bodies, in every infusion of plant and animal, be it open or covered, cold or warm' (L. Oken, *Die Zeugung* (1805) [Weimar: Hermann Böhlaus, 2007], 3). The term, which was introduced in 1763 by Martin Frobenius Ledermüller, 'referred to all microscopic animals' which until then had been called 'animalculae' (F. Vienne, 'Organic Molecules, Parasites, *Urthiere*. The Controversial Nature of Spermatic Animals, 1749–1841', in S. Lettow (ed.), *Reproduction, Race and Gender. Philosophy and the Early Life Sciences in Context* [Albany, NY: SUNY, 2014], 61). Authors such as Georges Canguilhem and François Jacob have discussed Oken's theory in the context of the history of the cell theory. Cf. Vienne, 'Organic Molecules'.

29   Oken, *Die Zeugung*, 144.

30   The polyp became a common point of reference in scientific debates on generation after Abraham Trembley's experiments. When Trembley discovered that polyps can multiply by being cut into parts so that each part forms a whole living being, he called this process 'reproduction' in order to distinguish it from preformationist understandings of 'generation'. Although Trembley highlighted the fact that he had discovered a certain mode through which new organic beings come into existence, 'well into the nineteenth century regeneration was the main sense in which living beings were "reproduced"' (Hopwood, 'Keywords', 292).

31   Oken, *Die Zeugung*, 155.

32   Ibid., 195. Of course, Oken concedes the existence of women. In his view, 'in the animal, the male must reproduce himself within a female, and the female was a hybrid, a combination of male and female. This meant that the female had the tendency to transmute the male animalcules into a female form, because the formative process took place in the womb. Therefore, the production of male or female was dependent on the female principle' (P. H. Reill, 'The Scientific Construction of Gender and Generation in the German Late Enlightenment and in German Romantic *Naturphilosophie*', in S. Lettow [ed.], *Reproduction, Race and Gender. Philosophy and the Early Life Sciences in Context* [Albany, NY: SUNY, 2005], 77) that could be a stronger or weaker within the female. If both principles – male and female – 'were of equal power, then the woman would produce a male because "her inborn masculinity would tip the balance"' (ibid.).

33 G. W. F. Hegel, *Philosophy of Nature. Encyclopedia II*, trans. A. V. Miller (Oxford: Clarendon Press, 2004), §353: 358.
34 Ibid., §348: 343.
35 Ibid., §345: 312.
36 A. V. Miller translates 'the passive moment'. The original reads: 'das Indifferente'.
37 Ibid., §369: 413
38 Ibid., § 369: 413.
39 'It is notable,' as Alison Stone writes, 'that in his 1821 *Philosophy of Right*, Hegel appeals to his philosophy of nature to support his claims about the proper social roles of men and women. Infamously maintaining that women's place is in the home ... he argues that it is through this gendered division of social roles that "The *natural* determinacy of the two sexes acquires an intellectual and ethical significance".' (A. Stone, 'Sexual Polarity in Schelling and Hegel', in S. Lettow (ed.), *Reproduction, Race and Gender. Philosophy and the Early Life Sciences in Context* [Albany, NY: SUNY, 2014], 266-7.) In addition, Hegel articulates his Eurocentric view of world history through the distinction between indifference and difference, claiming that the mental and spiritual characteristics of Africans correspond to 'the compact, differenceless mass of the African continent' whereas in Asia Spirit separates itself from Nature, and finally, in 'the Caucasian race' Spirit 'wrests itself free from the fluctuation between one extreme and the other' so that it becomes capable of creating world history. G. W. F. Hegel, *Philosophy of Mind. Encyclopedia III*, trans. W. Wallace and A. V. Miller (Oxford: Clarendon Press, 1971), §393: 43-4.
40 Schelling, *First Outline*, 231.
41 I. H. Grant, *Philosophies of Nature after Schelling* (London: Continuum, 2006), 137. He continues, 'The grounds of this extension lie in the consequences of the removal of a priori somatism from the genetic problem. Thus asked to comment on the prospects of a "threefold parallelism" between the developmental stages in organization (ontogeny), series of organizations (phylogeny) and earth history (geogeny) by Karl Joseph Windischmann, Kielmeyer responds favourably to the idea of thus extending the *Stufenfolge* through inorganic nature ... while remaining sceptical concerning the prospects for the execution of such programme' (ibid., 140).
42 Indeed, the then 'emerging sense that the earth and its inhabitants have an extended history' (Bowler, 'Theories of Generation', 96) was not closely linked to the various theories on generation and reproduction, and thus to attempts to understand the specificity of organic nature.
43 La Mettrie, quoted in Bowler, 'Theories of Generation', 89.
44 I. Kant, *Critique of Judgement*, trans. J. C. Meredith (Oxford: Oxford University Press, 2008), §80: 248.
45 Ibid., 247.
46 Ibid., 248.
47 Plato, *Timaeus*, trans. R.G. Bury (Cambridge, MA: Harvard University Press, 1975).
48 Ibid., 52d.
49 Stone, 'Sexual Polarity', 265.
50 F. W. J. Schelling, *Philosophical Investigations into the Essence of Human Freedom*, trans. J. Love and J. Schmidt (Albany: SUNY Press, 2000), 67-8.
51 F. W. J. Schelling, *Von der Weltseele - Eine Hypothese der Höhern Physik zur Erklärung des allgemeinen Organismus* (Stuttgart: Frommann-Holzboog, 2006), 255.
52 Stefani Engelstein ('Sexual Division and New Mythology') shows that Schelling, in his mythological construction, evacuates the positive feminine connotations from the

materials he reads. 'Although still called "mother", Schelling's ground is no longer a positive path towards which salvation strives, but rather a negative impetus away from which enlightenment travels. He even goes so far as repeatedly to identify the word most firmly associated with the feminine in Böhme and the Gnostics, namely "primeval [uranfängliche] wisdom"... with the masculine understanding [Verstand] which is God's contribution to the chaos of the Ground'.

53 Zammito, *Gestation*, 249.
54 Grant, *Philosophies of Nature*, 121.
55 Reill, *Vitalizing Nature*, 191.
56 'In this phase of the discussion generation and gender were vaguely linked, and when done, gender played a bit part in the drama that captured the imagination of educated Europeans' (ibid., 221).
57 See, e.g. L. Schiebinger, *The Mind has No Sex? Women in the Origins of Modern Science* (Cambridge, MA: Harvard University Press, 1991); C. Honegger, *Die Ordnung der Geschlechter. Die Wissenschaften von Menschen und das Weib* (Frankfurt/Main: Campus, 1991).

Chapter 13

MECHANICS BEYOND THE MACHINE IN KIELMEYER AND ESCHENMAYER

Jocelyn Holland

This essay proposes a new way of looking at the topic of mechanics in the writing of Carl Friedrich Kielmeyer. Such a focus goes against the grain of the established scholarship on Kielmeyer in two ways: in its emphasis on the mechanical over an organic theory that is certainly much more dominant in Kielmeyer's intellectual trajectory, but also in its re-scripting of the narrative established by Dorothea Kuhn, Thomas Bach and others who have paid attention to Kielmeyer's mentions of clockwork, machinery and related words. With reference to the work Kielmeyer is best known for, the 1793 Speech, Bach calls *Maschine* and *Maschinerie* a metaphor, and he is not incorrect.[1] At the same time, in doing so he establishes a dichotomy between a *Maschine* that is 'real' as opposed to 'not real' (because it is an organism), an opposition that is not a heuristically-useful tool of interpretation because it only tells half the story. To acquire a more complete picture, one needs to consider, in addition to the Speech, a much less well-known document where Kielmeyer's thinking about mechanical terms does not align with such earlier comments. This short piece is a letter Kielmeyer wrote to A. C. A. Eschenmayer after reading Eschenmayer's[2] 1797 work, *Propositions from Nature Metaphysics Applied to Chemical and Medicinal Subjects* [Säze aus der Naturmetaphysik auf chemische und medizinische Gegenstände angewandt].[3] The editor of the 1938 Kielmeyer edition dated this document to 1799, and gave it the title 'On Mechanical, Chemical, and Organic Movements', for reasons that will become clear below.

As John Zammito has documented in *The Gestation of German Biology*, Eschenmayer's writings played a pivotal role in the broader context of an approach to *Naturphilosophie*, a branch of philosophy that emerged during the 1790s and is often associated with the work of F. W. J. Schelling. Generally speaking, proponents of *Naturphilosophie* thought to observe affinities between laws of nature and aspects of the human psyche. Eschenmayer's 1797 publication of the *Propositions* was followed the very next year by a treatise devoted to developing laws of magnetic phenomena from nature-metaphysical propositions – 'one of the texts upon which

Schelling would draw for his own *Naturphilosophie* and, with Eschenmayer and others, build into a movement'.[4] The degree to which Kielmeyer himself adhered to or challenged the foundations of nature-philosophical principles is not clear cut, however. When another student of Kielmeyer's, Christian Heinrich Pfaff, published his memoirs he included this comment about his former professor's nature-philosophical leanings: 'Yet I cannot conceal the fact that Kielmeyer allowed himself not infrequently to be carried along to audacious hypotheses, and that he granted certain analogies too much importance,' adding 'the last comment can be applied to the explanation of the development of organic beings according to the laws of magnetism.'[5] In Kielmeyer's 1799 letter to Eschenmayer on the *Propositions*, however, he takes a more critical tone; for example, in his opening comment that he is largely in agreement with the 'end results' of Eschenmayer's investigation, but without being able to 'find himself' in Eschenmayer's chain of reasoning (*GS*, 42).

One of the goals of the present essay is therefore to use the 1799 document as a way of better understanding Kielmeyer's thinking about mechanics, and also a way of learning more about his position regarding Eschenmayer's nature philosophy. This, too, presents a challenge: just as Kielmeyer has trouble 'finding himself' in Eschenmayer's text, readers familiar with the *Propositions* might have difficulty finding Eschenmayer in Kielmeyer's critique. One reason for this difficulty is that Kielmeyer orders his remarks according to three kinds of movement (*Bewegung*), even if this term is not particularly foregrounded in Eschenmayer's *Propositions*. And when Kielmeyer does refer to one of the figures of central importance both for the *Propositions* and for other writings by Eschenmayer, the mechanical lever, he does something quite unusual by disintegrating it into a pile of little levers using an additional mechanical device – the wedge – which is not part of Eschenmayer's toolkit. Readers of the 1799 document are therefore confronted with a somewhat artificial conceptual order on the one hand and an unexpected pile of mechanical parts on the other.

In order to clarify the connection between the 1799 document and Kielmeyer's speech from 1793, the following discussion will show how in both cases the concept of relation (*Verhältniß*) is paramount, despite differences in the respective discursive environments. Whereas the 1793 Speech is concerned with the relations of various organic 'forces', in the 1799 document, 'relation' is attached to the conceptual apparatus surrounding the mechanical lever. That the lever can function as a figure of ratio or proportion is intuitive enough when its law of equilibrium is taken into consideration, and, as will be described below, Kielmeyer instrumentalizes the notion of mechanically-inflected relations in his reply to Eschenmayer.[6] In doing so, he constructs a mechanical figure that can be used heuristically, one which can also be understood as a more flexible alternative to the metaphor of machinery in the 1793 Speech.

The argument will be developed as follows: I begin with a brief overview of Eschenmayer's *Propositions* for those who might not be familiar with this work (also to underscore the idiosyncratic nature of Kielmeyer's response). The essay will then reconstruct Kielmeyer's categories of mechanical and chemical movement to explain how he connects them to the lever in a way distinct from Eschenmayer, before showing the relevance of this discussion to the problem of the organic. Although Eschenmayer, for his part, is quite content to explain organic movement in terms of the lever, Kielmeyer shifts away from an overt use of the lever at this point, and he is also quite careful to distinguish organic movement from the chemical and the mechanical. At the same time, I will argue that there is now a mechanical thinking operative in Kielmeyer's letter to Eschenmayer, and that such thinking impacts on how he articulates organic movement as well. As stated above, the larger goal is a new perspective on how Kielmeyer's mechanically-minded thinking develops through the 1790s and, in particular, how such thinking connects to emergent *Naturphilosophie*.

Eschenmayer's Propositions (1797)

For years, until a very recent upsurge of interest in *Naturphilosophie*,[7] Eschenmayer has been relegated to footnotes in studies devoted to those philosophers whose writings he most rigorously engaged, including Kant, Fichte and Schelling. One finds, with little exception,[8] only scattered references to him in scholarship on the Early German Romantics, who appreciated Eschenmayer's willingness to use metaphysical arguments in pursuit of connections between chemistry, the study of electricity and medicine. Eschenmayer himself envisioned his *Propositions* as a guidepost for future thinking about how to apply nature-philosophical ideas more broadly, and he divided his treatise into propositions relating to chemistry, mechanics and pathology. As one would expect, the most significant statements on the lever occur in the section on mechanics, but mechanical equilibrium is a constant theme throughout. Eschenmayer also makes a case in the preface for the exemplarity of the lever as defined by its ability to transcend simple categorical definitions. For example, citing the concept of irritability or excitability [*Reizbarkeit*], one of the central phenomena in the discussion of organic life around 1800,[9] he suggests that the reason why this concept was even introduced in the first place was because of the perceived difficulty in bringing animal movement under the same laws as the mechanical ones found in the theory of the lever: 'One does not consider,' he writes, 'that these laws are themselves *applied* and stand under the condition of even higher laws, which metaphysics demonstrates.'[10] These statements from Eschenmayer's mechanics in a nature-philosophical context are important with regard to Kielmeyer: they offer a way of working with mechanics without subscribing to the notion of an organic machine, much less a mechanistic philosophy.

Given that the lever, and the concept of equilibrium, are able to act as a conceptual bridge between the fields of mechanics, chemistry and medicine, it is difficult to over-emphasize the centrality of the lever in Eschenmayer's nature-philosophical thinking.[11] Even concepts and physical scenarios which, at first glance, would seem to have little to do with the mechanical theory of the lever are drawn into its orbit. To give just one brief example, consider the case of water temperature: 'Every temperature of water between its boiling and freezing points

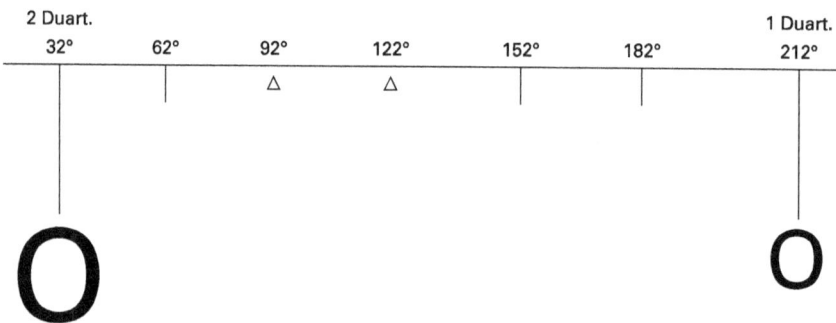

Figure 1 (From *Propositions*, 27.).
On the left of the diagram, one can imagine a quantity equivalent to two quarts of water at 32° F (the freezing point). On the right, there is one quart of water at 212° F (the boiling point). The mathematical average of 32° F and 212° F is 122° F, but the diagram illustrates how, when these two quantities of water are mixed, then, according to the law of the lever, because of the difference in volume, the middle temperature will actually be 92° F, which is where the fulcrum point is marked on the diagram.

can be understood as having emerged from two different temperatures, of which the one is larger, the other smaller, than the middle temperature.'[12] Because every temperature can be understood as a composite of the weight of the water and the 'degree of elasticity of its warmth', it can therefore, 'according to the analogy with the lever, be called a quantity of movement, and the middle temperature can be seen as a common hypomochlion [or fulcrum point], against which two such quantities of movement are working'.[13] If one recalls the law associated with the mechanical lever that says in the case of equal weights and velocities that the distance from the fulcrum point must also be the same, then, Eschenmayer argues, it must also be true that in the case of equal masses of water, the negative and positive degrees of elasticity are also in equilibrium: 'thus the mechanical law of the lever can be applied precisely to dynamic quantities'.[14] Eschenmayer provides a diagram to illustrate his point (see Figure 1).

What, then, would Kielmeyer have learned from Eschenmayer's writing on the lever and the diagrams he uses to articulate his ideas? Diagrams such as this one do not depict a particular object or piece of laboratory equipment. They visualize a philosophical idea as well as a theoretical outcome. In somewhat different language, we could say that they also have a function akin to the tendency of Blumenberg's absolute metaphors to act as a 'model' and 'point of orientation' that guides our thinking and 'gives structure to a world'.[15] Once the basic law of the lever is understood, one can apply it even in contexts not usually associated with levers. According to Eschenmayer's *Propositions*, the lever functions as an instrument of reason when we can use it to develop and visualize particular intuitions we might have about physical processes and then verify whether or not these intuitions are correct. His comments about the lever can also be understood within a broader tendency toward simplification, but analogical thinking is only useful in this sense when the same observations or intuitions can be re-applied productively in different contexts. The quantification of nature and natural processes is important in this regard, and Eschenmayer writes that it is 'an essential advantage of the dynamic way of looking at things that it excludes the multiplicity of specifically different materials and views the entire manifold of nature in terms of degrees'.[16]

Certainly, the figure of the mechanical lever retains its usefulness in the context of dynamic philosophy. Eschenmayer continues: 'Indeed, hereupon rests the place of mediation [*Vermittlungsort*], in which the most contradictory opinions of the physicists can be compared as well as the propositions of metaphysics for all possible hypotheses which the investigator of nature desires to establish.'[17] According to the theoretical scenario described, one could think of this 'place of mediation' as a fulcrum point that balances various intellectual 'forces', such as physics and metaphysics. If so, then the lever is more than a mere model or example: it arguably functions as one of the governing structural ideas of nature philosophy itself. At the end of the paper I will address the question of to what degree Kielmeyer might have internalized this idea.

Kielmeyer on Mechanical and Chemical Movement

As mentioned above, Kielmeyer's decision to use the concept of movement as the centrepiece of his letter to Eschenmayer seems at first glance somewhat unusual, given that this term is not particularly foregrounded in the *Propositions*. When movement does make an appearance in the *Propositions*, it is used to underscore that organic motion can be completely understood in mechanical terms within a nature-philosophical framework. Eschenmayer refers specifically to the lever to make this claim, as does Kielmeyer in his response. He even goes a step beyond Eschenmayer by focusing on two circumstances in which mechanical movement occurs: when the physical lever moves as a whole, and when it breaks down into parts.

The first circumstance describes the entire lever and can be represented as shown in Figure 2.

Unlike Eschenmayer, for whom the lever was more of a mathematical object used to illustrate nature-philosophical ideas, Kielmeyer is interested in the problem of material constraints.[18] This requires a different conceptual language, and Kielmeyer introduces the idea of cohesion into his model – first in terms of a *Zusammenhaltende Kraft* and later as *Cohäsion*. One of the standard late-eighteenth

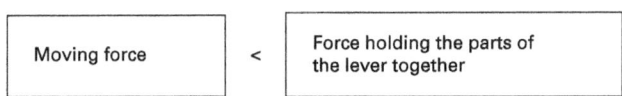

Figure 2.

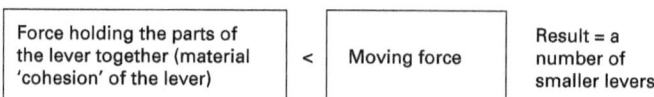

Figure 3.

century scientific reference works, Gehler's *Physical Dictionary*, gives us the following definition of cohesion, which he takes from the 1793 edition of Friedrich Gren's *Outline of the Theory of Nature*. According to the dictionary definition, the 'phenomenon of cohesion', which arises from the inherent 'force' of coherence, is such that the parts of each body are arranged in such a way that an external force is necessary to shift or separate them; degrees of cohesion account for divisions into solid and liquid, hard and soft, flexible and rigid bodies.[19] Note that Kielmeyer defines the 'whole' lever when it is at rest in similar terms to the lever in movement, with a focus on its material integrity. From this point of view, he writes that equilibrium is possible when 'the sum of the opposed motive forces is less than the force of cohesion' (*GS*, 43). (It seems an obvious statement. Clearly, the lever has to be strong enough not to break – there is a reason to construct them of wood or metal as opposed to straw.) What happens when those conditions are not met? In that case, the lever is destroyed [*zertrümmert*]. For Kielmeyer, this destruction occurs in a special way: into a number (in his words: an '*amas*' and '*Summe*') of tiny levers (*GS*, 43). This is the second case for mechanical movement, and can be approximated as shown in Figure 3.

If the force of cohesion that guarantees the material integrity of the lever is less than the force moving it, Kielmeyer adds, 'then instead of a movement of the lever as a whole there follows a movement of the sum of levers of which it is comprised, a destruction of the physical lever into levers, as if by a wedge, which acts as a lever on a collection of levers' (*GS*, 43). Note Kielmeyer's emphasis on the materiality of the lever as a physical (as opposed to mathematical or conceptual) object when he continues by stating 'a general condition for the physical lever to operate' is

> the inequality or relative weakness of two conflicting forces, one of which is the force of which the lever is comprised, through which it coheres in its parts, and generally *is*, and the other force, through which it is provoked to be effective as a lever. It is the same, whether the one or the other of these forces is greater, insofar as the effectiveness of the physical lever in general is not thus determined, but rather just the effect of the physical lever as the operation of a whole as opposed to the operation of a pile of several levers.
>
> *GS*, 43

Kielmeyer departs significantly from Eschenmayer both through his insistence on the physicality of the lever (the term 'physical' is entirely absent from the *Propositions*) and in the matter of scale. Within any given context, Eschenmayer's levers function on a single scale; they possess none of the self-similarity of parts and whole that defines Kielmeyer's description, where a single lever can be broken into parts which are themselves functioning levers, an operation that could then presumably be repeated over and over again on increasingly smaller scales. In other words: levers are constitutive of both parts and the whole of a dynamic physical model as described by Kielmeyer.

To understand more clearly what a fractured lever can do for Kielmeyer, however, the concept of chemical movement needs to be addressed. He describes two conceptual scenarios in which chemical movement can occur: the first is under circumstances of *Solution*, a term synonymous with *Auflösung*, or dissolution. This occurs when heterogeneous materials are joined into a homogeneous mass (such as when water bonds with sugar) and can be represented as follows (see Figure 4), using Kielmeyer's own terminology.

According to Kielmeyer, this scenario corresponds to the mechanical model associated with the destruction of the physical lever, where the lever represents the integrity of an individual substance before it is dissolved into solutes by binding with a solvent. The comparison with mechanical movement functions inasmuch as the solutes, like tiny levers, remain available. The lever operates in a grey area between model and material, as the simplest representation of a breakable thing that exists under a sheer tension of forces.

The second case in which 'chemical movement' occurs is during precipitation [*Präzipitation*], when a previously homogeneous mass divides into heterogeneous materials. Again, using Kielmeyer's own terminology, precipitation can be represented as such (see Figure 5).

This scenario corresponds to the other example with the lever: 'chemical movement in precipitation occurs under precisely the condition, under which the movement of the lever as a whole occurs. Cohesion overcomes [*überwältigt*] the force that is making or has made a violent incursion [*Attentat*] on the connection [*Zusammenhang*] of parts.' (*GS*, 45)[20]

Kielmeyer's conclusion from the comparison of chemical and mechanical movement is easily summarized. In each case, movement – as opposed to the stasis or *Ruhe* of relative equilibrium – occurs due to an inequality of two forces. Kielmeyer extrapolates from this idea to claim that such inequality is the condition for all movement in nature. This is an idea sympathetic to similar claims made in Schelling's *Naturphilosophie*, but the care Kielmeyer takes to appropriate the lever

Force of cohesion/ Affinity of component parts	<	The motive forces effecting a destruction of parts	First example of chemical movement

Figure 4.

The forces of cohesion unique to the individual materials	>	Those forces that are holding the individual materials in a homogeneous mass	Second example of chemical movement

Figure 5.

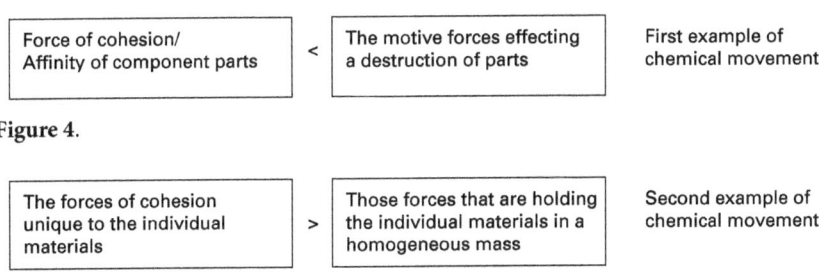

to describe the motion and rest of solid and liquid substances also raises the question of to what degree one can witness in the 1799 document the emergence of an explanatory template which could, retroactively, be held up alongside the descriptions of comparable types of motion made in the 1793 Speech.

To approach this question, one could first consider how Kielmeyer uses Eschenmayer's nature-philosophical lever to make a particular point. The lever is useful, he writes, because when we compare this mechanical instrument with the phenomena of chemical affinity, it is revealed to us how slight the difference between the foundations and causes of mechanical and chemical movement really is.

> The phenomena of adhesion, since not even fluid bodies can be compelled to that state, demonstrate all of the effects of physical levers and are at the same time incipient phenomena of affinity, the first stages of affinity. [The phenomena of adhesion] are the bridge from mechanism into the realm of chemistry... The bridge is furthermore indistinguishable from both areas.
>
> GS, 46

The lever therefore serves multiple functions for Kielmeyer. It operates within a mechanical context as an example for the requirements of movement, but it also transcends the same context in order to operate as a bridge figure that translates back and forth between the chemical and the mechanical. Is the lever acting as a metaphor in these contexts, the way the word 'machine' does in Kielmeyer's 1793 Speech? The answer is no: the functioning of mechanical levers is used to model certain types of motion and stasis, just as they are in Eschenmayer. And as with any sound model, this particular one also becomes a basis for further thought experiments, as witnessed by the fact that Kielmeyer places tentative constraints of materiality on Eschenmayer's immaterial levers. This particular use of the lever has a historical precedent in the years around 1800. One finds it in the work of Hegel somewhat later than Kielmeyer's reception of Eschenmayer, when Hegel uses the mechanics of the lever to structure a relationship of equivalence between the real and ideal in his definition of *Aufhebung*, but we can track a comparable usage already in the notes and fragments of Early German Romanticism, where the lever is used to mediate between concepts of quite different discursive origin, such as in Schlegel's statement that 'Act and hypothesis form a lever, belief is the hypomochlion.'[21]

Kielmeyer's attempts to define the common basis of mechanical and chemical phenomena lead him on a path closer to Eschenmayer's own chain of reasoning than he has been willing to admit. Like Eschenmayer, Kielmeyer is able to use Kant in order to think through the problem of why all chemical phenomena are accompanied by mechanical phenomena, when the converse is not necessarily true. Kantian nature-metaphysics tells us that matter is constituted through forces of expansion and contraction. In Kielmeyer's formulation, it is a single force which can operate in two opposing directions: 'the difference in the result of chemical and mechanical movement can be explained by the various degree of the identical forces that come into collision, with matter as the product of forces' (*GS*, 46). Despite the fine-tuning of differences between mechanical and chemical movement, Kielmeyer ultimately reaches the conclusion that it is not possible to speak of the priority of one or the other. Let us consider, for a moment or two, how the corresponding idea plays out in Eschenmayer's *Propositions* and in the broader context of German nature philosophy. The section in Eschenmayer's treatise devoted to chemical propositions takes the position that chemistry should strengthen its connection to dynamics, beginning with the basic idea that 'the existence of matter can be thought simply through the assumption of two original forces – these forces are the forces of attraction and repulsion'.[22] This idea is central to Eschenmayer's nature philosophy, and Jörg Jantzen and others have described how both he and Schelling understood all materials of nature as the quantified modification of these forces, extending over a scale that reaches from $1/\infty$ (a maximum limit for the force of attraction) to ∞ (a maximum limit for the force of repulsion).[23]

In his *Propositions*, Eschenmayer encourages us to think of matter in terms of degrees, such that 'a degree of matter' would be 'a quantity of the relationship in which the forces of attraction and repulsion stand toward each other'.[24] The various possible ratios of these two forces, quantified onto a scale, would then correlate to all the different materials in existence.[25]

To illustrate his point, Eschenmayer provides his readers with a diagram (see Figure 6).

In the diagram, A = the force of attraction and B = the force of repulsion.[26] Each member of the series represents a different quantitative possibility that, in turn, corresponds ideally to a different material found in nature, ideas that can be traced back to a dispute between Eschenmayer and Schelling on the construction of 'qualities'. Jantzen, for example, connects Schelling's Kantian-influenced understanding of qualities as 'various modifications of the basic forces [i.e. of attraction and repulsion J.H.]' to Eschenmayer's diagram, which is, according to Jantzen, an attempt to encompass Schelling's idea 'through a kind of mathematical progression, which in particular expresses *gradation*'.[27] Jantzen also reminds us that, already in the *First Outline of a System of Nature Philosophy* from 1799, Schelling expressed his dissatisfaction with mathematical representations of quality.[28]

Eschenmayer has no such qualms, however: even in his later work, he will continue to use quantitative thinking, no matter how speculative the context. In the *Propositions*, we can see that he keeps the ideas underlying this diagram as well as other quantitative figures as a point of reference, even as he shifts his focus from chemical to mechanical propositions.

If it is the business of dynamics to develop the concept of matter with regard to the category of quality, and if the application of those propositions to chemistry

$$+n \quad +3 \quad +2 \quad +1 \quad 0 \quad -1 \quad -2 \quad -3 \quad \quad -n$$
$$A.\ B, \ldots A.\ B, A.\ B, A.\ B, M.\ A.\ B, A.\ B, A.\ B, \ldots A.\ B.$$

Figure 6 (From *Propositions*, 12.).

showed us that this science has to do with degrees of reality, then it is left to the business of mechanics to demonstrate these degrees of reality in relation to one another in order, where possible, to find the laws of their equilibrium.[29]

Chemical affinity, then, should be understood as the process through which matter, in its qualitative relations (as expressed by Figure 6) strives for equilibrium. According to Eschenmayer, the laws of mechanics can be applied to this same theory when we understand mechanical equilibrium as the opposition of two 'degrees of reality'[30], an idea he also illustrates with reference to the mechanical lever.

This additional glimpse into the workings of Eschenmayer's *Propositions* is significant, with regard to Kielmeyer, for several reasons. One can observe, for example, that the groundwork for one of the key ideas Pfaff disparages Kielmeyer for embracing – the reduction of living phenomena to the laws of polarity – is already diagrammed in Eschenmayer's *Propositions*. The passages above from the *Propositions* also illustrate the way in which a mechanical figure like the lever might be operating covertly as well as openly within Kielmeyer's thinking about mechanical and chemical movement. Eschenmayer and Kielmeyer share the idea that the different results of chemical and mechanical movement can be explained from the varying degrees of forces associated with matter as a product of forces in collision. And both apply the laws of mechanics to chemical affinity.

The Limits of the Organic

The question remains, to what degree a mechanical model remains useful for Kielmeyer when the third kind of movement, organic movement, is considered. This is a bigger question than might appear to be the case at first glance, because it connects to a range of topics in Kielmeyer's writing that cannot be easily grouped under a single umbrella, including what it means for something to be assimilated in a physiological sense during organic processes. One could also consider the ways in which ideas connected to the lever's conceptual apparatus (equilibrium, ratio and translation) continue to play a role in Kielmeyer's thinking about organic movement.

Without going so far as to want to 'melt' chemical and organic movement together, as he claims Eschenmayer does, Kielmeyer begins by acknowledging similarities. Both chemical and organic products, for example, emerge within conditions of fluidity coupled by a force of expansion. And just as chemical movement homogenizes and heterogenizes, the organism does as well: a base assimilates an acid, water assimilates salt, just as animals and plants assimilate their own nutrients (GS, 51). All this aside, Kielmeyer is much more interested in emphasizing the striking differences between chemical and organic processes. One example he provides is 'neutralization'. In chemical movement, heterogeneous materials can become completely neutralized: when an acid assimilates a base, or a base assimilates an acid, the product is a chemically neutral substance that is neither acidic nor basic. This phenomenon is already indicated by the etymology of the neutral: *ne utrum*, neither the one nor the other. The organism that digests its food, however, does not receive qualities from it; nor does it become neutralized in the process: the individual who 'humanifies' (*vermenscht*) an ox through the process of digestion, does not 'oxify' (*verochst*) at the same time (GS, 46). Also humorous – though perhaps unintentionally – are Kielmeyer's remarks about borderline cases, such as Greenlanders and Tierra del Fuegans who have been observed to adopt something of the seal nature due to the monotony of their diet, or about the loving relations of married couples who become similar to one another through a process of neutralizing and mutual consumption, but for Kielmeyer these are the exceptions that prove the rule. Regardless how seriously one takes Kielmeyer's examples, his argument about neutralization is important because it introduces a structural asymmetry into the discussion. The examples of mechanical and chemical movement Kielmeyer provides emphasize their reciprocity, as a relationship between two objects in collision, a tension of forces or the interaction of chemically diverse substances.

Such asymmetry becomes even more pronounced when Kielmeyer considers the limits of organic assimilation, an idea that is connected to the concept of a chemical saturation point. In chemical reactions, the saturation point is one of maximum concentration, when a solvent is not able to dissolve any more of a substance without forming a precipitate. For Kielmeyer, this point is not completely transferable, due to the insatiability of organic assimilation, both in a material sense and in an intellectual or spiritual one.

> Just as its intellectual egoism [*geistiger Egoismus*] relates and transforms everything to itself, so too its material individuality seeks to appropriate and homogenize the entire world to itself.... In every plan there either is or desires to be a world-conqueror Alexander [the Great] ... What material individuality does not accomplish, intellectual egoism does, where the material egoism ends, the intellectual one begins – and where the world ends for intellectual egoism, it creates for itself a new one, in whose formation and desire the individual lives itself out [*verlebt*].
>
> GS, 52–3[31]

It is a remarkable passage, not least of all for suggesting that one of the defining features of organic movement is, when pushed to a particular limit, the instrument of its extinction. And more precisely: it is in the transition from material egoism to 'intellectual egoism' that the organism – namely, the human organism – incurs this risk.

The relationship between material and intellectual egoism is an intricate one for Kielmeyer. It is one of supplementarity, as the passage just cited suggests, but Kielmeyer also states that the faculty forming the basis of each of them is identical 'to the degree that a comparison is allowed'.

> The intellectual egoism [*Der geistige Egoismus*] homogenizes and heterogenizes incessantly through wit and acumen and reproduces through fantasy, the

material individuality of the organism [does so] through the faculties of assimilation, secretion, and reproduction ... The organic individual certainly has a maximum in its development, but the saturation point of the species lies in infinity, the species is immortal.

GS, 53

The idea of an 'intellectual' (or 'spiritual') egoism has a certain resonance with philosophies of the subject, although the particular term *geistiger Egoismus* is uncommon around 1800. Even if any reference to external sources for this concept would be a bit tenuous, there are nevertheless at least two possible connections to the 1793 Speech on the relations among organic forces. The first relates to a passage from the 'Urfassung' of the Speech not included in the first published edition. In this passage, in order to emphasize the interconnectedness of the organic world across scale – from organs to increasingly complex systems of individuals, species and beyond – Kielmeyer adopts the voice of nature itself, speaking in the first person. This ego is supremely aware of its creative and destructive capabilities: 'I have already on occasion annihilated stars above ... I have exterminated animal species from the earth – but what happened: a different species arose.' (p. 47)

In the final pages of the published Speech, Kielmeyer makes another clear connection between the production of mind in the human sphere and the regeneration of bodies in the animal world: 'With every privation that the human spirit experiences on one side, it supplements itself, like the polyp, on the other.' (p. 45) Passages such as these convey a dual sense of equilibrium: as in nature, the human spirit restores balance by compensating itself for a loss. At the same time, the concept of disequilibrium where organic movement is concerned is already firmly in place in 1793. It arises, for example, in the question of 'as to whether here and there an over-preponderance in the destructive forces does not actually arise, and whether species whose forces of preservation are inferior must not eventually themselves pass away [*sich...verleben müßten*]' (p. 44). It is well known that humans excel at such destruction: 'What so many higher animals are capable of when it comes to some mussel species, humanity and its species are capable of doing with respect to so many other, and even higher animals.' (p. 44) Such augmentation, and imbalance, of power manifests itself as much through an expansion of the capabilities of the self as it does through a destruction of the environment: 'With the rationality that turned up in humanity's organisation, humankind obtained the capacity to freely alter (within certain limits) the relation of the other forces that it has in common with the other animals. It created microscopes and telescopes for eye and ear, and thereby heightened its sensorial capacity, and who knows whether further similar improvements may not be applied to smell and touch.' (p. 44)

In short, Kielmeyer's 1793 Speech introduces the idea of disequilibrium-states in the organic world, but frames them within a system of checks and balances. The equilibrium of the organic world is a dynamic equilibrium, one of constant ruptures and compensations. The 1799 document, by contrast, takes up the asymmetrical patterns of such disequilibrium-states without the corresponding language of correction and compensation. The position Kielmeyer occupies when describing organic movement around 1799 is unique: distinct from his earlier writing *and* Eschenmayer's *Propositions*, while at the same time indebted to each of them. Eschenmayer, who already established MnE = nME as a viable formula for equilibrium-states (where M = mass, E = elasticity, and n is a numerical coefficient), sees no problem applying this law 'to the animal machine': 'The insight of this law, that n M E = M n E, makes animal movement comprehensible to us and its explanation, which we otherwise sought beyond the sphere of mechanics, still lies within it for the most part' (Eschenmayer *Sätze*, 52). Without having to commit ourselves to a particular terminology, such as *Lebensgeist, Nervenfluidum, Lebensprincip*, etc., Eschenmayer suggests that we can simply assert that such a material exists, because, without it, animal movement would not be comprehensible: the fluid environment acts as a medium for the forces of expansion and contraction. One can quibble with Eschenmayer's logic, but the important point is that he reaches to the same equation of mechanical equilibrium associated with the lever, while Kielmeyer, who has diligently allowed the lever to play a role in his descriptions of mechanical and chemical movement, no longer mentions the lever when it comes to organic movement.

Kielmeyer's response to Eschenmayer is an extremely useful document for Kielmeyer-studies as a whole. Provoked to respond to Eschenmayer's blend of mechanical models and nature-philosophical thinking, Kielmeyer reveals the degree to which he is willing to adapt some ideas and concepts while distancing himself from others. The emphasis on mechanics has the additional advantage of casting Kielmeyer's earlier work in a somewhat different light, as we consider the tension between equilibrium- and disequilibrium-states in the Speech on the relation of organic forces. To take a less tentative and – in the spirit of Kielmeyer's day – somewhat more speculative approach, however, one could ask why the lever disappears in Kielmeyer's discussion of organic movement. A possible answer lies in Kielmeyer's idea that a species' immortality is one of unbridled excess. In the 1799 document, Kielmeyer makes this statement without any of the modifying forces found in the speech from 1793. What if we were to consider this idea also from a mechanical perspective, just as we have been doing with the other ideas from the 1799 document? The caveat is that a different aspect of the mechanical lever is being articulated: one which distances itself from a counterbalance of forces while approaching the notion of an Archimedean point, as an extreme expression of agency. Such extreme expressions of agency emerge around 1800 in the writings of the Early German Romantics, where levers without arms are fantasized and the human is described as an all-powerful point. This comparison would need to take into consideration distinctions Kielmeyer maintains between individual and species, but the idea of an unbounded productive potential remains the same. On the positive side, it becomes possible to explain why the lever is no long mentioned, namely, because its instrumental value has been reconfigured. No longer a heuristic tool of analogy, it has become a blind spot for having been internalized into the dynamic operations of intellectual individual productivity in the service of the species.

Notes

1. T. Bach, *Biologie und Philosophie bei C. F. Kielmeyer und F. W. J. Schelling* (Stuttgart: Frommann-Holzboog, 2001), 90–1 and passim.
2. For information on Eschenmayer, readers can consult the entry on him in *The Bloomsbury Dictionary of Eighteenth-century Philosophers*. As regards his relation to Kielmeyer, it is important to note that Eschenmayer first attended the *Karlsschule* in Stuttgart, then studied medicine in Tübingen (1793 to 1796) before becoming a student for a 'short time' in Göttingen; he then took a position as an assistant medical doctor in 1797 (V. Abaschnik, 'Eschenmayer, Adolph Karl August', in the *Bloomsbury Dictionary of Eighteenth-Century German Philosophers*, ed. H. F. Klemme and M. Kuehn [London: Bloomsbury, 2016], 194).
3. Susanne Lettow, in her essay 'Generation, Genealogy, and Time: The Concept of Reproduction from *Histoire naturelle* to *Naturphilosophie*,' makes brief mention of this document, but situates it within the historical narrative of reproduction she establishes in her essay. For Lettow, the document is of interest because of the analogy Kielmeyer makes between the 'articulation of reproduction' and 'the narrative of creation' (in S. Lettow (ed.), *Reproduction, Race and Gender in Philosophy and the Early Life Sciences* [Albany, New York: SUNY, 2014], 32).
4. J. H. Zammito, *The Gestation of German Biology* (Chicago: University of Chicago Press, 2017), 301.
5. C. H. Pfaff, *Lebenserinnerungen* (Kiel: Schwers'sche Buchhandlung, 1854), 41.
6. Like a simple mechanical scale, a lever is in equilibrium when the product of the weight and the weight's distance from the fulcrum on one side is equal to the product of a second weight applied to the lever and *its* distance from the fulcrum on the other side. This is with reference to 'first class levers', where the fulcrum point lies in between the two weights.
7. At the forefront of this trend are Daniel Whistler and Benjamin Berger. Their volume, *The Schelling-Eschenmayer Controversy, 1801*, forthcoming in 2020 with Edinburgh University Press, sets a much-needed milestone in the scholarship on Eschenmayer.
8. In addition to Bach and Kuhn (see below), one can consult the chapter on Kielmeyer in R. J. Richards' *The Romantic Conception of Life: Science and Philosophy in the Age of Goethe*, Chicago: University of Chicago Press, 2002), 238–51.
9. For a quick overview of the competing perspectives and suggestions for further reading, see the entry on Christoph Wilhelm Friedrich Hufeland in *The Bloomsbury Dictionary of Eighteenth-century German Philosophers*, eds. Heiner F. Klemme and Manfred Kuehn (London: Bloomsbury, 2010), 362–5.
10. A. C. A. Eschenmayer, *Sätze aus der Natur-Metaphysik, auf chemische und medicinische Gegenstände angewendt* (Tübingen, 1797), xiii.
11. Ziche has also noted that Eschenmayer turns often to the lever in his discussion of chemical and magnetic phenomena and that it has achieved 'the role of a general illustration of the Eschenmayerian method of construction'. P. Ziche, *Mathematische und naturwissenschaftliche Modelle in der Philosophie Schellings und Hegels* (Stuttgart: Frommann-Holzboog, 1996), 214.
12. Eschenmayer, *Sätze*, xiii.
13. Ibid.
14. Ibid., 26.
15. Blumenberg, *Paradigms for a Metaphorology*, trans. Robert Savage (Ithaca, NY: Cornell University Press, 2010), 22.

16 Eschenmayer, *Säze*, 44.
17 Ibid.
18 Note that Kielmeyer's willingness to think in terms of materiality, at least hypothetically, is quite different from the state of affairs as described by Iain Hamilton Grant in the 1793 speech: 'Since no particular body, macrobody (substrate, primal substance, aether, etc.) or microbody (protoplasm, intermaxillary bone, etc.), provides the "standard of measurement" or "basal element" of organic nature, the forces that define its specificity do not inhere in an isolably organic nature' I. H, Grant, *Philosophies of Nature after Schelling* (London: Continuum, 2006), 136.
19 See "Cohäsion" in J. S. T. Gehler's *Physikalisches Wörterbuch oder Versuch einer Erklärung der vornehmsten Begriffe der Kunstwörter der Naturlehre*, vol. 5 (Leipzig: Schwickert, 1799), 197.
20 This is a statement that contains more than a hint of violence in the original German, given that *Attentat* is also the word used to describe an assassination.
21 '*Der Act und die Hypothese* bilden einen *Hebel*, der *Glaube* ist das Hypomochlion.' F. Schlegel, *Kritische Friedrich-Schlegel-Ausgabe*, ed. Ernst Behler (Munich: Schöningh, 1958—), -), 18, 404.1002.
22 Eschenmayer, *Säze*, 2.
23 See J. Jantzen, 'Adolph Karl August von Eschenmayer', in T. Bach and O. Breidbach (eds), *Naturphilosophie nach Schelling* (Stuttgart: Frommann-Holzboog, 2005), 155. Jantzen quotes Schelling's 1797 *Ideen zu einer Philosophie der Natur*, where Schelling writes that '*alle* Qualitäten' are to be observed 'nur als verschiedne Modifikationen und Verhältnisse der Grundkräfte.' Jantzen accidentally omits the words 'und Verhältnisse' from his quote, but for our purposes they are of the utmost importance: it is precisely the notions of ratio and relation that lay the groundwork for a productive comparison with the lever.
24 Eschenmayer, *Säze*, 37.
25 See also the letter from Eschenmayer to Schelling written in Kirchheim, 21 July 1801, in *Aus Schellings Leben* vol. 1: 1775-1803, ed. G. L. Plitt (Leipzig: S. Hirzel, 1869), 336-43.
26 In *Mathematische und naturwissenschaftliche Modelle*, Ziche shows how the concept of 'line' is used to model the concepts of identity and difference for both Schelling and Hegel (200). In particular, he describes how the 'line of cohesion' (*Kohäsionslinie*) in Schelling's work was applied to both the magnet and the mechanical lever. Ziche does not refer to Eschenmayer in this context.
27 J. Jantzen, 'Eschenmayer und Schelling. Die Philosophie in ihrem Übergang zur Nichtphilosophie' in W. Jaeschke (ed.), *Religionsphilosophie und spekulative Theologie: Die Streit um die Göttliche Dinge (1799-1812)* (Hamburg: Felix Meiner, 1994), 74-5.
28 Ibid., 76.
29 Eschenmayer, *Säze*, 21.
30 Ibid., 22.
31 In German, the full passage reads as follows: 'Wie sein geistiger Egoismus alles auf sich bezieht und in sich verwandelt, so sucht seine materiale Individualität die ganze Welt sich anzueignen und sich zu homogenisieren. In jeder Pflanze strebt oder ist ein Welteroberer *Alexander*, dem nach dem Mond gelüstet, wach; was die materiale Individualität nicht vollbringt, vollbringt der geistige Egoismus, wo jener aufhört, fängt dieser an – und wo diesem die Welt aufhört, schafft er sich selbst eine neue, in deren Ausbildung und Begehren sich das Individuum verlebt'.

Chapter 14

THE LOGIC OF ORGANIC FORCES: HEGEL'S CRITIQUE OF KIELMEYER

Benjamin Berger

Life is a profoundly important category in Hegel's system. Although he insists that inorganic phenomena are more fundamental or basic features of the natural world, living things are, for Hegel, more fully expressive of being itself, since they are self-determining individuals. And despite the fact that living things only achieve maximal self-determination in the human *spirit*, the organic realm is nevertheless essential to the ontological progression that begins with space, time and matter and culminates in the 'absolute spirit' of art, religion and philosophy: organic life – and animal life in particular – constitutes the most rationally developed stage of nature, and it paves the way for what is to come in the spiritual life of humanity.

Central to Hegel's account of animal life is a rational derivation of the biological functions that allow an organism to maintain itself and its species.[1] Those functions are nothing other than the three forces elaborated upon by Kielmeyer in his 1793 Speech. It is not clear whether Hegel ever read Kielmeyer, but he was certainly familiar with the theory of organic forces discussed in the Speech.[2] In fact, Hegel was familiar enough with the theory and its influence to develop a substantial criticism of it in the *Phenomenology of Spirit*. Indeed, despite employing the same tripartite system of sensibility, irritability and reproduction in his own account of life, Hegel rejects Kielmeyer's understanding of the rational connection between the forces.

This paper explores Hegel's logic of organic functions and how it relates to his critique of Kielmeyer. The paper proceeds in four sections. First, I seek to elucidate Hegel's critique of Kielmeyer in the *Phenomenology* – a critique which already points towards Hegel's mature understanding of the organic forces. In the second section, I turn to Hegel's explicit account of the organic forces as discussed in the *Science of Logic* and *Encyclopaedia* Philosophy of Nature. That Hegel appears to be drawing upon a broadly Kielmeyerian schema in both the *Logic* and Philosophy of Nature raises questions about the supposedly non-empirical justification of his logic of organic functions. These questions are addressed in the third section of the paper. Finally, in the fourth section, I consider the way that Hegel's distinctive conception of speculative organics leads him to understand the history of life as devoid of rational significance – putting him at odds, once more, with Kielmeyer. I conclude by suggesting that Kielmeyer's interest in the rational laws governing the development of life *as a whole* – and not merely the life of the *individual* and its species – ought to inspire an exploration of alternatives to Hegel's ahistorical philosophy of nature.

(I) Pre-Hegelian Philosophy of Nature as Observing Reason

Hegel's sustained critique of Kielmeyer is found in the chapter on 'Reason' in the *Phenomenology of Spirit*, specifically within the section on 'Observing Reason: Observation of Nature'. In this section, Hegel seeks to demonstrate the manner in which a certain form of reasoning about nature – namely, reasoning about nature via experimentation and observation – fails to elucidate the being of nature. Thus, the critique of Kielmeyer plays a role in Hegel's more general attempt to understand the way that certain forms of reason fall short of their own goal, namely, to grasp, in thought, the rational structure of reality.

Once we have made our way to the chapter on reason in the *Phenomenology*, consciousness knows itself to be identical to everything that is. That is to say, the form of consciousness that is rational does not oppose itself to an object that is *other* than it – and which can therefore be determined as either more or less real than consciousness; instead, rational consciousness or reason knows *everything* to be rational, mind and world alike. In the most basic form of reasoning, however, it is the world, or nature, that becomes the site of grasping the rational structure of reality. Yet unlike more logically primitive forms of knowing, such as perception, reason does not passively wait for the natural world to make its mark upon the mind and thereby make itself known. On the contrary, Hegel says, 'reason sets to work to know the truth' by '[making] its own observations and experiments'.[3]

The most abstract form of such observation and experimentation yields nothing more than descriptions of natural phenomena. However, since such description can go on *ad infinitum*, universal, rational truth is nowhere to be found.[4] Observing reason then attempts to describe only those characteristics of natural phenomena that appear to be *essential*.[5] Yet here again, grasping the rationality of nature proves to be a difficult task: the characteristics that appear to be 'essential' to observing reason also appear to be unstable. For instance, the organism that is observed to be the organism it is on account of its claws and teeth can lose these naturally-occurring tools and thus, apparently, lose its essential being.[6] To seek the universal, *unchanging* truth of natural phenomena, observing reason falls back upon an idea that had occupied a more basic form of consciousness in the *Phenomenology*, namely, the understanding, which seeks to comprehend the *laws* of nature. Yet, as Stephen Houlgate makes clear, whereas the understanding regards the laws of nature as 'purely intelligible', reason regards these laws as *observable*.[7] And it is in this way, Hegel argues, that observing reason seeks to grasp the universal, rational truth in particular phenomena: by observing the laws of nature immanently at work in the sensible.[8]

After considering forms of observing reason that attend to inorganic nature,[9] the relationship between the inorganic and organic (a consideration that includes Hegel's critique of Treviranus),[10] and external teleology (which interprets environmental conditions as existing *for the sake of* the organism),[11] Hegel goes on to consider two far more complicated – and, apparently, more rational – ways to observe reason in nature. The first of these corresponds to Kielmeyer's theory of organic forces, and the second corresponds to Steffens' research on the history of the earth. That Hegel focuses on both Kielmeyer and Steffens in this section is noteworthy, since, as Cinzia Ferrini argues, it can be read as a critique of the Schellingian school of nature-philosophy – Kielmeyer being one of its original inspirations and Steffens its most enthusiastic advocate.[12] It will be important to keep in mind, however, that, although Hegel is critical of this general movement in the philosophy of nature, he does not reject it entirely. As Robert Stern writes, 'Hegel's aim is to show that while we must respect the achievements of the natural sciences, we should not exaggerate them.'[13] Thus, Hegel's critique of Kielmeyer does not amount to an outright dismissal of his thought but rather a criticism of the idea that it can make fully intelligible the rationality at work in the organic world.

Kielmeyer raises three questions in his 1793 Speech on organic forces. First, he asks, which forces are to be found in the greatest number of living things? In other words, what are the forces that animate the 'great machine of the organic world'? Second, what varying proportions of these forces can be seen to correspond to the variety of species in the organic world, and what laws explain these varying proportions and thereby distinguish one species from another? And third, how does knowledge of these forces ground an understanding of the development of life as a whole? (p. 31) The first of these questions is quickly answered in the Speech. Kielmeyer identifies sensibility, irritability, the reproductive force, the force of secretion and the force of propulsion as the five most general forces in the organic world. Hegel is primarily interested in Kielmeyer's answer to this question and the second, that is, the question about the proportions of force and the laws that govern the differences between species. This should not be surprising, given the fact that these are the issues Kielmeyer himself spends the most time discussing in the Speech. It is worth noting, however, that the third question – the question that potentially pertains to the historical development of life on earth – is *not* of particular interest to Hegel (although he does remark upon the misguided nature of this general manner of thinking).[14] I will return to this point in the final section below.

On Hegel's interpretation of Kielmeyer, the organic forces constitute the inner aspect of life, in contradistinction to life's outer aspect, namely, the material systems which express the organic forces. And since Hegel takes his cue from Kielmeyer in focusing primarily upon the first three forces, he goes on to consider the three (or, more precisely, four) physiological systems that make manifest the forces of sensibility, irritability and reproduction. The nervous system gives expression to the force of sensibility; the muscular system gives expression to the force of irritability; and the digestive and reproductive systems give expression to reproductive force, since the nutritive process allows the organism to reproduce itself as the individual organism *it* is, and procreation allows the organism to reproduce itself as *another* individual of the same kind.[15] Insofar as Kielmeyer's theory about the forces is exemplary of 'observing reason', it claims that both the inner force and the material system that expresses this force can be observed.[16] Importantly, though, the law that *unites* the inner and the outer *cannot* be observed, and this, according to Hegel, is where Kielmeyer's theory of organic forces begins to break down:[17] there can be no laws regulating the relationship between the inner and the outer aspects of life, because life is – according to observing reason itself – something that is *fundamentally* unified, and empirical laws necessarily treat their objects as being constituted by antithetical elements.[18]

The law that regulates the relationship between the inner force and the 'outer', material system is not the only law that Hegel believes goes unexplained in Kielmeyer's theory. More importantly, perhaps, Kielmeyer attempts and fails to articulate rational laws that govern the relationships *between* the organic forces. The so-called laws of compensation aim to describe the existence of the diverse forms of life on earth by referring to the way any of the three forces is expressed at a determinate degree of intensity depending upon the intensity of the other forces – an increase in one of the forces corresponding to a decrease in another. (Or, in some cases, the increase in one *aspect* of a force is related to a decrease in another aspect of the *same* force) (see pp. 41–2). For example, species characterized by a great deal of sensibility – i.e. life-forms that *feel more* – have a decreased force of reproduction: mammalian forms of life have less reproductive power than vegetal life, which is barely sentient (p. 41). In seeking to observe laws governing the proportions of organic forces expressed in different species, however, observing reason destroys the possibility of ever grasping the nature of life. And, according to Hegel, there are a few connected reasons as to why this is the case.

First, observing reason understands life – rightly, Hegel thinks – to be intrinsically *unified*. Indeed, this form of consciousness is particularly interested in the nature of *life*, since life appears to be – like reason itself – unified in its difference from itself: consciousness is rational, because it knows itself to be at home in a world to which it is not simply identical; and the living thing, likewise, appears to be at home in an environment from which it differs. Moreover, a living thing appears to be unified insofar as it is an internally differentiated and therefore meaningfully organized body. Thus, observing reason – and this includes Kielmeyer's theory of the organic forces – seems to grasp life as a unified whole. Yet in seeking to explain the unity of life with reference to forces that are taken to be *essentially different* from one another, observing reason begins to contradict its own understanding of the unity of life. And this is because, as soon as one *begins* with different forces, all attempts to *unify* those forces will be derivative of their original separation. If we observe living things in front of us that appear to be organic wholes, and if, in order to explain the activities of these individuals, we construct laws referring to *originally different* forces, then our theory of life will necessary become mechanistic, treating each of the organic forces as fundamentally external to the others. As Hegel puts it, 'the organism is apprehended' here as 'a dead existence; [the] moments [or activities of the organism] so taken pertain to anatomy and the corpse, not to . . . the *living* organism.'[19]

What would it mean to apprehend the organism as *living*? We will not be in a position to answer that question until we turn to Hegel's own account of the organic forces. But it is going to have something to do with making more explicit the implicit teleology in Kielmeyerian thought; that is, it will have something to do with making fully transparent the way that the organic forces are intrinsically united as different means of maintaining the life of the individual organism and its species. Before we turn to Hegel's own account of life and his transformation of Kielmeyer's theory of organic forces, however, it is important to note a number of further problems that Hegel identifies in this theory.

We might think that, because Hegel criticizes Kielmeyer's treatment of the organic forces as originally *different* from one another, Hegel himself might go on to argue that there is *no* difference between the forces. But this is not what Hegel ultimately believes. In fact, Hegel seeks to demonstrate that presupposing the forces of life to be *originally* different from one another actually blinds observing reason to their genuine difference and the way that a living organism is *actually* differentiated according to its various biological functions. According to Hegel, if sensibility, irritability and reproductive force are understood to be originally or immediately different from one another – external to one another in the way that mere things are outside of one another in space – then the difference between these forces will only ever be grasped as a *quantitative* difference; their difference 'becomes one of magnitude, and [thus] there arise laws of the kind, for example, that sensibility and irritability stand in an inverse ratio of their magnitude, so that as the one increases the other decreases'.[20] In this way, Hegel claims, the Kielmeyerian loses sight not only of the *unity* of life but its *qualitatively differentiated* character. As Andrea Gambarotto writes, the 'qualitative difference [of organic functions] is ... *reduced* to a difference in magnitude'.[21] By reducing organic processes to proportions of force – that is, by reducing quality to quantity – observing reason once again fails to hit its mark. What is more, this reduction of qualitative difference to difference in intensity is connected to Kielmeyer's failure to provide an adequate account of the fundamental unity of the organic forces. Indeed, according to Hegel, it will only be through a consideration of the qualitative differentiation of the organism's activities that a philosophy of nature will be capable of demonstrating the unity of life.

On a Hegelian account, part of the problem with Kielmeyer's approach is that it seeks to discover rational laws where there are none. Attempting to discover laws that explain the diversity of species is a fool's errand, for Hegel, because the diversity of species *is not rationally determined*; the differences between various forms of life are 'exposed to contingency'.[22] Perhaps an astute observer of nature will be capable of documenting the different levels of sensibility, irritability and reproduction exhibited by a given form of life, but these differences have nothing to do with anything rational; on the contrary, 'what characterises their sensuous being is just this, that they ... manifest the freedom of nature released from the control of the concept ... irrationally playing up and down the scale of contingent magnitude'.[23] I will return to this point below and suggest that, despite Hegel's compelling criticisms, the early Kielmeyer offers us a promising way forward in the philosophy of nature, precisely insofar as he makes a space for a rational, philosophical account of the (historical) differentiation of life. What we learn here, however, is that part of Hegel's critique of Kielmeyer's theory of organic forces concerns the idea that this theory seeks rationality where there *isn't* any. Or, to be more precise: this theory seeks too much reason in nature; for nature, properly understood, is *self-external* reason, a rational form of existence in which there is a proliferation of contingency and *irrationality*. Thus, according to Hegel, the Kielmeyerian mistakenly employs mathematical laws to explain the empirical differences between forms of life, because she has a faulty understanding of nature as straightforwardly or immediately rational.

At the end of the section on the observation of nature in the *Phenomenology*, we learn that reason cannot in fact discover rationality in empirical laws that immanently govern organic life – or anything in nature for that matter. As Hegel writes, 'even if reason digs into the very entrails of things and opens every vein in them so that it may gush forth to meet itself, it will not attain this joy'.[24] And this is because observing reason tries to *observe* itself, i.e. to observe rationality, in something that is *given* to it. We are not dealing with the primitive form of knowing that Hegel calls 'sense-certainty', which takes *what is* to be that which is *immediately* given as 'this'; nor are we dealing with 'perception', which takes *what is* to be that which is perceived as a thing with properties. Observing reason is far more rational than these forms of knowing. Nevertheless, there is an irreducible element of empiricism in the observation of nature. Kielmeyer seeks to grasp the laws of organic forces through *observation*, not through any sort of *a priori*, rational derivation. Indeed, this is why Kielmeyer begins with the forces *as they are given*: sensibility appears to be different from irritability, our experiments reveal a negative correlation between them, and on this basis we construct empirical laws of organic nature (pp. 32–3). Yet in doing so, we have turned life into the non-living; we have intellectually dissected the organism by setting its various activities side by side and putting them into a strictly quantitative relationship to one another.

It follows, according to Hegel, that even if Kielmeyer does observe patterns in nature, these patterns do not explain anything about the rational structure of life. The conclusion of Hegel's critique of Kielmeyer's theory of forces, therefore, is that it simply is not rational enough, for it does not make explicit the rationality inherent in life – the most rational of natural phenomena. This may appear to be an odd claim to make given the suggestion above that – for Hegel – Kielmeyer tries to provide a rational explanation of *too many* things, such as the differences between the variety of species. But the latter issue – i.e. Kielmeyer's apparently hyperrational attempt to explain the diversity of species in terms of mathematical laws – is symptomatic of this more general problem with Kielmeyer: he lacks a *truly* rational approach to understanding nature, i.e. a rational approach that knows its limits and focuses exclusively on that which is *logically necessary* (as opposed to those empirical 'contingencies' that Kielmeyer seeks to comprehend in terms of proportions of force). The naïvely optimistic search for a plenitude of rationality in nature, then, depends upon an *empiricism* that undercuts the very project of observing *reason* in nature. For the observer of nature, 'the truth of the law is found in *experience* ... but if the law does not have its truth in the *concept*' – that is, if the law does not have its truth in *logic* – 'it is a *contingency*, not a necessity, not, in fact, a law'.[25] A truly rational comprehension of life, for Hegel, will be one that explains the logically necessary connections between the essential activities of organic life, articulating the genuine unity *and* difference between these activities. But this means that a properly rational philosophy of nature must leave behind the rational methods of observation and experimentation. At least, so it seems.

(II) Hegel's Philosophy of the Organism

It is important to keep in mind that, by the time we come to the chapter on observing reason in the *Phenomenology*, consciousness has already proven itself to be *rational*; this is a form of consciousness that knows itself to be at home in the world and thus discovers important truths about the nature of reality. This should be taken as a clue regarding Hegel's ultimate assessment of Kielmeyer and the other 'observers of nature' criticized in the *Phenomenology*: their philosophies of nature are indeed rational, even if they are not *fully* rational. Seen from this perspective, it is not surprising that, despite Hegel's immanent critique of Kielmeyer's observational philosophy of nature, he also finds something true in Kielmeyer's conception of the organic forces. Indeed, Hegel's rationalist system will integrate Kielmeyer's empirical discoveries. I therefore turn now to Hegel's own account of life in order to see how the idea of the organic forces is incorporated, and significantly revised, in Hegel's system.

In the chapter on observing reason in the *Phenomenology*, Hegel actually tells us quite a lot about how a proper philosophy of life might proceed. I have sought to bracket those remarks, in part, because they are philosophically unjustified from Hegel's own perspective. Yet these remarks do point ahead to what Hegel will go on to develop in his later philosophy of nature, the mature version of which was outlined in the 1817 *Encyclopaedia* and revised in 1827 and 1830.[26] Indeed, already in 1807, Hegel argues that a truly rational comprehension of life and its activities will require the philosopher to step back from experience and unfold – in sheer thought – the way that more complicated organic activities logically emerge from more basic activities.[27] The task of a proper philosophy of life is therefore to understand the way that life necessarily differentiates itself as a series of logically distinct moments. In this way, the 'forces' of life become logical – but no less real – 'moments' of a living process.

In both the *Science of Logic* and the *Encyclopaedia* Philosophy of Nature, the living individual is characterized by three functions: sensibility, irritability and reproduction. Yet rather than simply observing that these functions are different and then attempting to integrate them into a unity with reference to mathematical laws governing their intensity, Hegel seeks to show how sensibility logically gives rise to irritability and how irritability logically gives rise to reproduction. To understand this logical development, we need to begin by noting that, at this point in the dialectic of both the *Logic* and the Philosophy of Nature, Hegel has already derived the necessity of *life*; that is, he has argued that it is *rationally necessary* that there exists something that remains what it is *through* difference or otherness.[28] The most basic form of life, for Hegel, is the living individual. Although he will ultimately go on to discuss the individual's relationship to its species, the individual is – logically speaking – more basic; indeed, the very existence of the species depends upon the activity of the individual organism.

At the most basic level, a living individual is a being defined by its inner determinacy and the immediate experience of *itself*, which proves its individual character. This immediate experience of self, i.e. the 'undivided identity of the subject with itself',[29] is nothing other than self-feeling or sensibility. Because the individual is, at this basic level, *itself*, its vitality is expressed as a 'purely internal vibration',[30] i.e. an utterly *inward* experience. Incidentally, this explains why Hegel views plants as not fully expressive of life, for they do not possess the inwardness of sensibility, the most basic sense in which life achieves individuality or selfhood.[31]

Implicit in this moment of sensibility is the fact that there is a real *difference* between the individual organism and its environment. This is why living individuals do not merely '[receive] all externality into [themselves], while reducing it to the perfect simplicity of selfsame universality';[32] rather, Hegel says, living things *posit* themselves as *different* from their surroundings by acting in the world, by *moving* in one way or another and thus distinguishing themselves from their environment. Hegel understands this moment of differentiation as the 'opening up of the negativity that is locked up in simple self-feeling, or is an ideal, not yet a real, determinateness in [sensibility]'.[33] In other words, there is a feature of organic life that is logically implicit in sensibility but does not express itself as sheer sensibility. And this 'negativity' is *irritability*: the organism's capacity to move. In the Philosophy of Nature, Hegel also identifies this second feature of organic life as the 'general standpoint of reality' and 'the general standpoint of matter',[34] since it makes fully explicit the worldly character of being alive.

This moment of negativity is crucial, for Hegel, and it is something that can only be understood if one attends to the logical development of the idea of an individual life.[35] Yet the negativity of irritability does not exhaust the activity that is logically implicit in the life of the individual. Since we are unpacking the logical structure of a living *individual* – a living individual that *is what it is*, or *becomes* what it is, through *difference* – we know that the moment of negative differentiation cannot be the whole story. On the contrary, acting in a world that is *external* to the organism must serve the purpose of maintaining that individual's *selfhood*. Therefore, according to Hegel, irritability leads, logically, to reproduction.[36] The organism finds itself in a world that is *different* from it; the organism immediately feels itself as the individual it is, primarily through the experience of suffering (in the form of hunger, pain, etc.); the organism acts in the world in an attempt to alleviate its suffering; and by transforming substances of the external world into objects *for its own use*, the organism maintains itself, reproducing itself through its relationship to that which is different from or external to it.

This way of understanding reproduction is somewhat one-sided in that it emphasizes the organism's relationship to that which is *different* from it. For while life is, indeed, at home in *difference*, it is also *at home* in difference. Unpacking this latter aspect of life reveals that the individual organism does not only act upon an alien nature in order to return to itself in the nutritive process; the individual also acts in relation to other individuals *of the same species*. In sexual reproduction, the individual organism does not relate to a being from which it essentially *differs*; it relates to a different being to which it is essentially *identical*. The result of such an encounter is not the maintenance of the *individual's* life, but the life of the *species*. And this is significant, according to Hegel: in the perpetuation of the species through sexual reproduction, the individual gives itself over to a 'self' that is not the immediate self of the individual but the mediated self of the whole species;[37] in this way, life proves to be most fully united with itself through difference.[38]

This logical development is presented in both the *Logic* and the Philosophy of Nature – a point to which I return below. In the Philosophy of Nature, Hegel goes on to claim that these three functions of the organism 'have their reality in three systems, namely, the *nervous* system, the *circulatory* system, and the *digestive* system',[39] before further differentiating the structures of these systems. Thus, Hegel seeks to demonstrate that the functions of organic life are necessarily expressed in physiological systems that are, by definition, extended in space: sensibility, irritability and reproduction are embodied activities of life.

Hegel therefore seeks to show how these functions are necessary features of the organic world and that their differences from one another can only be understood by attending to the way the logic of life immanently generates them as *moments* of a more general vital process. Sensibility, irritability and reproduction are not, then, originally opposed forces; instead, they logically develop out of one another according to rational necessity, thereby constituting a logical series of qualitatively distinct functions. This logical necessity is precisely what is missing in the nature-philosophical research of someone like Kielmeyer, and this prevents him from grasping the unity of *and* difference between the organic 'forces'. Indeed, by failing to attend to the logical connections between the forces while nevertheless insisting that the laws of organic life can be rationally comprehended, thinkers such as Kielmeyer find themselves positing strictly quantitative laws, such as the laws of compensation, that treat the forces of life as originally different (i.e. not fundamentally unified) and yet only quantitatively distinguishable (i.e. without any determinate, qualitative difference).

(III) Reason and Experience

From a Hegelian perspective, Kielmeyer understands the forces of life to be originally distinct from one another, since that is how they appear in experience. When one simply observes the life of an individual, one notices disparate activities. To discover their original unity would require one to retreat from conscious experience and enter into the rational or logical structure of life itself. It is perhaps significant that Hegel fails to consider Kielmeyer's own suggestion that there may be a single, *original* force from which the others are derived, as this implies that Kielmeyer himself was open to the possibility of understanding the organic forces as emergent from one another (p. 31).[40] On Hegel's reading, however, Kielmeyer is overly empirical in his attempt to grasp the structure of organic life and this blinds Kielmeyer to the original unity of life's activities. The crux of Hegel's criticism of Kielmeyer, then, ultimately concerns the proper methodology in the philosophy of nature. If we begin with observation, we will fail to grasp the unity of life; for it is only through the rational derivation of the logical moments of life that one achieves genuine knowledge of organic unity.

But have we not just seen that Hegel himself employs the concepts of sensibility, irritability and reproduction in his supposedly *a priori* philosophy of life? Does this not problematize Hegel's explicit aim to develop a strictly *rationalist* account of the organic realm? Are we meant to believe that he simply pulls these concepts out of thin air, rather than being influenced in some important respect by the natural sciences of the day? Surely it is no coincidence that Hegel's logic of organic functions mirrors Kielmeyerian organics in certain respects. Yet how can he, on the one hand, criticize the empiricism of someone like Kielmeyer, and on the other, appear to draw upon empirically-based research in developing a speculative account of the organism?

Insofar as we focus on Hegel's philosophy of nature, there are relatively straightforward answers to these questions, and these answers get to the heart of the distinctiveness of his nature-philosophical method. In the *Encyclopaedia* Philosophy of Nature, he makes the following remark:

> The Philosophy of Nature takes up the material which physics has prepared for it empirically, at the point to which physics has brought it, and reconstitutes it, so that experience is not its final warrant and base. Physics must therefore work into the hands of philosophy, in order that the latter may translate into the concept the abstract universal transmitted to it, by showing how this universal, as an intrinsically necessary whole, proceeds from the concept.[41]

And in the Introduction to the *Encyclopaedia* Logic, to which Hegel refers the reader, he writes:

> Thus, philosophy does owe its *development* to the empirical sciences, but it gives to their content the fully essential shape of the freedom of thinking (or of what is *a priori*) as well as the validation of *necessity* (instead of the content being warranted because it is simply found to be *present*, and because it is a fact of *experience*).[42]

In these passages, Hegel makes an important distinction between (i) the manner in which philosophers first come to comprehend the logic of nature and (ii) that logic itself as an entirely self-sufficient dialectic. Insofar as human beings have achieved a certain understanding about the world through the modern natural sciences, such empirical knowledge provides the philosopher of nature with the data necessary to uncover nature's inner logic. In this way, the philosophical system of nature is dependent in an important sense upon empirical knowledge, for that knowledge is necessary for the philosopher's system to be developed in the first place. The case of Kielmeyer and the observers of nature is exemplary here: it would not have been possible for Hegel to articulate the logic of the organic functions in the Philosophy of Nature had he not first become familiar with the generally Kielmeyerian system of sensibility, irritability and reproduction.[43]

But this does not mean that the system of nature as such, i.e. the intrinsic logic of nature – (ii) above – is *ontologically* dependent upon experience. On the contrary, the logical series presented in the *Encyclopaedia* is, from a philosophical perspective, a fully self-sufficient progression constituted by nature's intrinsic rationality. The logic of organic functions found in the Philosophy of Nature, therefore, stands on its own as a *rational* derivation of sensibility, irritability and reproduction. What the example of Kielmeyer teaches us, then, is that Hegel depends upon the natural sciences for a sufficient understanding of the natural world, and it is on this basis that he is positioned to uncover the strictly logical or rational structures at work in nature's atemporal, rational progression – a progression that begins with spatial extension and culminates in animal life. That Hegel relied upon the natural sciences when uncovering this progression for the first time does not make it any less rational – so long as that progression can be shown to proceed immanently, i.e. without reference to empirical justification.

When we turn to the *Logic* things are not as straightforward. One could argue that the *Logic* – like the Philosophy of Nature – draws upon empirical discovery in order to first identify the fundamental structures of reality, structures which are, in principle, intelligible without reference to experience. But Hegel does not understand the *Logic* to operate in this manner. What is more, if we were to read the *Logic* along these lines, it would lead to further hermeneutic challenges regarding the systematic differences between the *Logic* and the Philosophy of Nature: if both branches of philosophical knowledge depend upon the empirical sciences to first observe certain patterns in nature, what ultimately distinguishes logic from nature-philosophy?

Another way of reading Hegel on this issue would be to suggest that he has simply gotten ahead of himself at the end of the *Science of Logic* in seeking to derive the necessity of sensibility, irritability and reproduction in the life of an individual. That a logic of life appears in the system prior to the Philosophy of Nature is not necessarily a problem. Hegel is quite clear about the fact that the logic of life found in the *Logic* is not the logic of *natural* life or the life of *organisms* but rather the structure of any life whatsoever – be that life natural or spiritual. Hegel thus develops a compelling way to think about the difference between the abstract structure of life *in general* and the concrete structure of life found in the natural world. It is telling, however, that Hegel's own general logic of life includes, as an apparently necessary aspect of the dialectic, references to sensibility, irritability and reproduction; indeed, this suggests that Hegel's 'general' logic of life has strayed quite far into a concrete logic of natural life. For Hegel himself distinguishes the latter from the former on account of the fact that concrete, natural life involves a vital activity that takes place *within the external world*: natural organisms are extended in space, related to environments from which they differ, and take on a 'multiplicity of . . . formations' or 'shapes' which correspond to those environments.[44] In other words, it is only in concrete reality – and specifically in the natural world – that 'life' manifests itself in, and in relation to, external being. And given the fact that life must necessarily be related to an 'outside' in order to be characterized by the functions of sensibility, irritability and reproduction, it seems that an abstract or general logic of life ought not to refer to such concrete functions. It is very possible, then, that when Hegel begins to consider the shape of life and the moments of sensibility, irritability and reproduction in the *Logic*, he has exceeded the bounds of abstract thought and – without logical justification – entered prematurely into his nature-logic.

A thorough treatment of these issues pertaining to the different parts of Hegel's system is beyond the scope of this paper.[45] What should be clear at this point is that, in both parts of the system, Hegel's account of the organic functions is *meant* to be a strictly rational account, precisely because its philosophical justification depends exclusively upon logic (be this an abstract or concrete logic of life). For Hegel, the intelligible structure of life is made apparent by logic alone, even if that logic refers, in certain ways, to experience. With this in mind, we can go on to consider an issue more directly related to the philosophical significance of Kielmeyer's Speech.

(IV) Kielmeyerian Organics After Hegel

Recall that Hegel understands the logic of life in terms of a living individual and, ultimately, that individual's relationship to its species. What Hegel does not make room for in either his abstract or concrete logic of life is a philosophical consideration of the structural relationships *between* various species. This does not mean that Hegel lacked curiosity about such relationships. In fact, he was deeply interested in comparative anatomy and had particular respect for the research of Cuvier, Kielmeyer's student.[46] But even this respect for the research of someone like Cuvier is not accompanied by an incorporation of that research into the rational progression that constitutes the philosophy of nature proper. And this is because the species compared by the anatomists are not determined, *as the species they are*, by any sort of rational necessity. On the contrary, according to Hegel, the 'impotence of the concept in nature generally, subjects not only the development of *individuals* to external contingencies ... but even the *genera* are completely subject to the changes of the external, universal life of nature.'[47] This is why it is misguided when observers of reason, such as the comparative anatomists, 'seek conceptual [i.e. rational] determinations everywhere'.[48]

There is certainly room in Hegel's system for a rational account of the structural relationships between very broad categories of life. Unlike the more abstract logic of life presented in the *Science of Logic*, the more concrete logic presented in the Philosophy of Nature involves a graduated sequence of life-forms: in his speculative organics, Hegel unpacks the logic that leads from the proto-life of the earth to the impoverished life of plants to the fully manifest vitality of animal existence. In this regard, Hegel does think it is important to rationally derive different forms or kinds of life. However, as Wendell Kisner writes, although 'in the *Philosophy of Nature* Hegel will attempt to show the relevance of the ontological determinacies of sensibility, irritability and reproduction/species to plant and animal life ... natural contingency rules out the possibility that, say, the scarab beetle *Popillia japonica* with its specific color pattern could be systematically derived.'[49] What is more, the specific *proportions* of the scarab beetle's organic functions cannot be derived: for Hegel, there is no fully rational means of arriving at knowledge of scarab life, and as the 'Observing Reason' chapter of the *Phenomenology* makes clear, one cannot grasp the necessary existence of a form of life by simply appealing to the laws of compensation.

Thus, although Hegel insists upon the philosophical significance of the structural relationship between vegetal and animal life *in general*, this is the furthest his logic of life will go in articulating ontological connections between different forms of life. And while Hegel may refer to the research of the comparative anatomists in his lectures on the philosophy of nature, this research – which concerns the genetic relationships *between* species – is not properly *philosophical* research on life, since it does not have anything to do with the immanent *logic* of life. That logic is characterized by concrete universality, which means that the essence of life, for Hegel, ultimately concerns the relationship between a universal kind and its instantiation, i.e. the species and the individual. The philosophy of life, in other words – and the philosophical significance of the organic functions – has to do with the way an individual achieves its true individuality (however temporarily) by giving itself over to its species.

Significantly, this focus on the species-individual relationship will yield no philosophical knowledge regarding the *individuation of the species*, that is, the event whereby a form of life is produced *in the first place*. Hegel's logic of life thus remains silent about the way a species comes into being and is set apart from other forms of life as the distinctive form of life it is. This brings us back to Kielmeyer and the third major question raised in the 1793 Speech: how does our knowledge of the organic forces help us to understand the development of life *as a whole*? Elsewhere, Kielmeyer remarks, 'The history of the phenomena yielded by our Earth as a whole must, according to the concept of natural history, address not only the question of their present state, but also that of the states preceding and perhaps succeeding the present one – thus, how it *is*, how it *was*, and how it *will be*' (p. 60). Despite Kielmeyer's subsequent reservations about the generation of novel life-forms throughout the history of the earth, there is clearly an impulse in the early Kielmeyer to discover the rational laws at work in the history of nature.[50] This impulse is by no means entirely unique to Kielmeyer, but it is nevertheless a significant aspect of his view about how those who study life and nature more generally ought to proceed. And there is undoubtedly something compelling about this perspective. To put it simply: should a speculative philosophy of nature not seek to explain something about the manner in which life *as a whole* transforms itself, i.e. how life determines itself as a *variety of kinds*? Indeed, how can the *being of life* be understood if the processes of speciation have been left outside nature-philosophical speculation?

It is well-known that Hegel was unconvinced of theories regarding the supposed transmutation of species. And I take it that this fact, on its own, is not particularly worthy of philosophical reflection. What *is* worth considering, however, is why Hegel thinks that, even if species *did* undergo metamorphosis and thereby produce new forms of life, the philosophy of nature *would not be required to attend to such events*. In fact, for Hegel, the philosopher of nature would be *led astray* in thinking about the history of life. When Hegel claims – at the end of his critique of the 'observers of nature' – that 'organic nature has no history',[51] and when he makes similar remarks about the earth in his speculative geology,[52] his point is not that nothing *happens* in nature, or that changes in nature do not occur. His point is that nothing of *philosophical significance* happens in nature. And this is because, according to Hegel, there are no *rational laws* motivating such changes. Whatever physical and chemical alterations the earth undergoes, whatever forms of life come into and go out of existence – for Hegel, these changes are contingent events. Perhaps empirical science can tell us something about contingencies of this kind, but the natural scientist's explanations will never be capable of being taken up by a *rational* philosophy of nature. The natural sciences will go on observing nature, claiming to have discovered rational laws where there are no truly rational laws to be found. To be sure, the rationality immanent to life 'exerts its influence', according to Hegel, 'but *only to a certain degree*'.[53] The philosophy of nature ought not to become distracted, therefore, by theories regarding the supposed 'origination of the more highly developed animal organisms from the lower'.[54]

Might there, however, be another mode of philosophizing about nature? Might there be a rationalist mode of investigating the essence of life that does not focus exclusively upon the logical connections between the individual organism and its species? That is, might there be a non-Hegelian form of speculation that could draw upon Kielmeyer's observations in order to discover the strictly rational laws that animate the *history* of the 'great machine of the organic world'? If John Zammito is right to suggest that 'in the address of 1793, Kielmeyer undertook what Arthur Lovejoy famously termed the "temporalization" of the "great chain of being"',[55] how might speculative philosophy do justice to the spirit of Kielmeyer's organics?

Given the fact that Hegel's argument against the idea that the history of nature is rational and of philosophical importance is to be found, in part, in the 'observing reason' chapter of the *Phenomenology* – and given the fact that the chapter on 'observing reason' can be read as Hegel's most detailed criticism of Schellingian philosophy of nature – it may be to *Schelling* that we ought to turn in order to discover a form of speculative thought that is attentive to nature's history.[56]

Notes

1 Given the way we tend to think about biological kinds today, I use the term 'species', as opposed to 'genus', throughout this paper.

2 Cf. T. Bach, 'Kielmeyer als "Vater der Naturphilosophie?" Anmerkungen zu seiner Rezeption im deutschen Idealismus' in *Philosophie des Organischen in der Goethezeit: Studien zu Werk und Wirkung des Naturforschers Carl Friedrich Kielmeyer (1765-1844)*, ed. K. T. Kanz (Stuttgart: Franz Steiner, 1994), 246, 248.
3 G. W. F. Hegel, *Werke in 20 Bänden* (Frankfurt am Main: Suhrkamp Verlag, 1970), vol. 3, 185-6; *Phenomenology of Spirit*, trans. A. V. Miller (Oxford: Clarendon Press, 1977), 145. Emphasis modified. Further references to Hegel's *Werke* abbreviated as *W* followed by volume number.
4 *W* 3: 188; *Phenomenology*, 148.
5 *W* 3: 189; *Phenomenology*, 148.
6 *W* 3: 190-2; *Phenomenology*, 149-50.
7 S. Houlgate, *Hegel's* Phenomenology of Spirit: *A Reader's Guide* (London: Bloomsbury, 2012), 125. Emphasis modified.
8 *W* 3: 194; *Phenomenology*, 152-3.
9 *W* 3: 193-5; *Phenomenology*, 152-4.
10 *W* 3: 196-8; *Phenomenology*, 154-6. Cf. Houlgate, *Hegel's* Phenomenology of Spirit, 126.
11 *W* 3: 198-9; *Phenomenology*, 156-7.
12 See Ferrini's discussion of 'Hegel's double move, to criticize the main "academic" line of reasoning of Schelling's school and to demote Schelling's own philosophy of nature, with its claims to speculative truth and to absolute knowing, to the phenomenological level of Observing Reason' (C. Ferrini, 'Reason Observing Nature' in *The Blackwell Guide to Hegel's* Phenomenology of Spirit, ed. Kenneth R. Westphal [Oxford: Wiley-Blackwell, 2009], 111). Ferrini is more precise than my comments above may suggest; she emphasizes the close relationship between Schelling and Steffens while recognizing Kielmeyer's independence as a thinker. Cf. 'Reason Observing Nature', 106-11. Although I do not address it here, Kielmeyer's difference from Schelling's speculative school of nature-philosophy is significant. As Gabrielle Bersier puts it, Kielmeyer was 'an empirical scientist' who employed a 'blend of observation and intuition' and yet was 'wary of his contemporaries' turn to extrasensory speculation' (G. Bersier, 'Visualising Carl Friedrich Kielmeyer's Organic Forces: Goethe's Morphology on the Threshold of Evolution' in *Monatshefte* 97.1 [2005], 20). That being said, Schelling's oft-quoted remark about Kielmeyer initiating a 'new epoch of natural history' is telling (F. W. J. Schelling, *Sämmtliche Werke* [Stuttgart and Augsberg: Cotta, 1856-61], vol. 2, 565). Despite his intellectual autonomy, Kielmeyer does unintentionally commence a distinctive nature-philosophical tradition. For the sake of this paper, what matters most is that *Hegel* understood Kielmeyer (and Herder) to be key inspirations for Schelling's development away from Kant and towards his own philosophy of nature. Cf. Bach, 'Kielmeyer als "Vater der Naturphilosophie?"', 248.
13 R. Stern, *Hegel and the* Phenomenology of Spirit (London: Routledge, 2002), 103.
14 *W* 3: 224-5; *Phenomenology*, 178. Cf. T. Bach, '"Aber die organische Natur hat keine Geschichte...": Hegel und die Naturgeschichte seiner Zeit' in *Geschichtlichkeit der Vernunft beim Jenaer Hegel*, ed. R. Beuthan (Heidelberg: Universitätsverlag, 2006).
15 *W* 3: 204; *Phenomenology*, 161.
16 *W* 3: 204-5; *Phenomenology*, 161-2.
17 *W* 3: 205; *Phenomenology*, 162.
18 *W* 3: 211; *Phenomenology*, 167.
19 *W* 3: 210; *Phenomenology*, 166. Emphasis modified.
20 *W* 3: 207; *Phenomenology*, 163. Emphasis modified.

21 A. Gambarotto, *Vital Forces, Teleology and Organization: Philosophy of Nature and the Rise of Biology in Germany* (Dordrecht: Springer, 2018), 120. Emphasis modified.
22 *W* 9: Addition to § 370, 503; *Philosophy of Nature, Part II of the Encyclopaedia of the Philosophical Sciences*, Second Edition, trans. A. V. Miller (Oxford: Oxford University Press, 2004), 418.
23 *W* 3: 109–10; *Phenomenology*, 165.
24 *W* 3: 186; *Phenomenology*, 146.
25 *W* 3: 192; *Phenomenology*, 151.
26 According to Hyppolite, the younger Hegel was more sympathetic to the theory of organic forces and 'in the *Realphilosophie* of 1803–4 and in that of 1805–6, Hegel especially emphasized these [organic] moments' (J. Hyppolite, *Genesis and Structure in Hegel's* Phenomenology of Spirit, trans. S. Cherniak and J. Heckman [Evanston: Northwestern University Press, 1974], 250). Yet as Olaf Breidbach points out, Hegel was already developing his mature position regarding organic nature in 1805 and thus 'breaks with the romantic categories of nature' even before developing his phenomenological critique of the observers of reason ('Das Organische in Hegels Jenaer Naturphilosophie' in *Hegels Jenaer Naturphilosophie*, ed. K. Vieweg [Munich: Wilhelm Fink, 1998], 318). See also Bach, 'Kielmeyer als "Vater der Naturphilosophie?"', 247n.
27 See, for example, *W* 3: 206, 212; *Phenomenology*, 163, 168.
28 *W* 6: 468; *Science of Logic*, p. 760. *W* 9: § 336, 334; *Philosophy of Nature*, 270.
29 *W* 9: § 353, 437; *Philosophy of Nature*, 357.
30 *W* 6: 478; *Science of Logic*, 768.
31 *W* 9: § 344, 374; *Philosophy of Nature*, 305. We can note a difference between Hegel and Kielmeyer on this point, as the latter remarks that traces of sensibility are to be found in plant life, even if these are only analogous to sensation proper (p. 34).
32 *W* 9: 478; *Science of Logic*, 768. Translation modified.
33 *W* 9: 478; *Science of Logic*, 768.
34 From the notes of Johann Rudolf Ringier in the Winter Term 1819/1820. Hegel, *Vorlesungen Über die Philosophie der Natur*, ed. W. Bonsiepen (Hamburg: Felix Meiner, 2012), 158.
35 It is worth emphasizing that Hegel's interest in tracing the different organic functions to their fundamental unity does not mean that he rejects the idea that the functions are antithetical *in some sense*. On the contrary, he argues that the genuine antitheses between the functions is *reduced to merely quantitative* antitheses by thinkers such as Kielmeyer who display a 'lack of acquaintance with the *logical* nature of these antitheses' (*W* 3: 208; *Phenomenology of Spirit*, 164). In other words, what the observers of nature fail to recognize is the *negativity* that truly differentiates the functions of an organism.
36 *W* 6: 479; *Science of Logic*, 769.
37 *W* 6: 483–4; *Science of Logic*, 772.
38 Reproduction turns out to be 'the reality and the basis' of sensibility and irritability (*W* 9: § 353, 437; *Philosophy of Nature*, 357), which explains how plants – despite being impoverished forms of life – are indeed alive; they simply lack the fully differentiated character of animal life in which sensibility and irritability come into their own (*W* 9: Addition to § 353, 438; *Philosophy of Nature*, 358).
39 *W* 9: § 354, 439–40; *Philosophy of Nature*, 359–60.
40 As Bach points out, Kielmeyer's suggestion about an 'original force' was particularly important for Schelling. Bach, 'Kielmeyer als "Vater der Naturphilosophie?"', 241.

41 *W* 9: Addition to § 246, 20; *Philosophy of Nature*, 10. Translation modified.
42 *W* 8: Remark to § 12, 58; *The Encyclopaedia Logic, Part I of the Encyclopaedia of the Philosophical Sciences*, trans. T. F. Geraets, W. A. Suchting and H. S. Harris (Indianapolis: Hackett, 1991), 37.
43 Kielmeyer is, of course, not the only thinker of the period to explain the distinctive features of life in terms of organic forces! And, as discussed above, it is not clear whether Hegel even had access to Kielmeyer's Speech. Nevertheless, it is Kielmeyer who popularizes the tripartite system consisting of sensibility, irritability and reproduction, and it is indeed this Herderian-Kielmeyerian system that Hegel adopts and significantly transforms. Moreover, according to Bach, Kielmeyer prepares the way for subsequent nature-philosophical considerations of life by emphasizing the way sensibility, irritability and reproduction work together in order to animate the life of an organism – a process that requires a higher level of philosophical understanding than merely anatomical description (Bach, 'Kielmeyer als "Vater der Naturphilosophie?"', 239–40). Finally, even if it were to turn out that Kielmeyer *in particular* had not been especially important for Hegel, the main point stands: Hegel employs the terminology that he does in these sections, because he is drawing upon the empirical sciences of his day – and this does not necessarily compromise his claims about the strictly rational justification of his conception of life.
44 *W* 6: 471; *Science of Logic*, 762.
45 For a more detailed and sympathetic discussion of these issues (and one that focuses on the logic of chemistry), see J. Burbidge, *Real Process: How Logic and Chemistry Combine in Hegel's Philosophy of Nature* (Toronto: University of Toronto Press, 1996).
46 See, for example, the Remark and Addition to § 368 in *W* 9: 500–16; *Philosophy of Nature*, 415–28.
47 *W* 9: Remark to § 368, 502; *Philosophy of Nature*, 416. Translation and emphasis modified.
48 *W* 9: Addition to § 368, 504; *Philosophy of Nature*, 418. Translation modified. Ferrini suggests that the mature Hegel is even more sympathetic to Cuvier, and she argues that, in Cuvier's empirical research, attention is paid to a form of necessity that Hegel understood to be consistent with his own logic of life. She writes, 'It seems to me that we have sufficient evidence to conclude that Hegel views Cuvier's law of the correlation of forms in organized beings as the modern empirical theory that conforms both to Aristotle's speculative principle and to the syllogistic linkage of the concept' (C. Ferrini, 'From Geological to Animal Nature in Hegel's Idea of Life' *Hegel-Studien* 44 [2009], 72).
49 W. Kisner, *Ecological Ethics and Living Subjectivity in Hegel's Logic: The Middle Voice of Autopoietic Life* (Hampshire: Palgrave Macmillan, 2014), 209.
50 See J. H. Zammito, *The Gestation of German Biology: Philosophy and Physiology from Stahl to Schelling* (Chicago and London: University of Chicago Press, 2017), 264.
51 *W* 3: 225; *Phenomenology*, 178.
52 See, for example, *W* 9: Addition to § 339, 348; *Philosophy of Nature*, 283.
53 *W* 9: Addition to § 368, 503; *Philosophy of Nature*, 418. Emphasis modified.
54 *W* 9: Remark to § 249, 31–2; *Philosophy of Nature*, 20. Emphasis modified.
55 Zammito, *Gestation*, 258.
56 See B. Berger, 'Schelling, Hegel, and the History of Nature' in *The Review of Metaphysics* 73 (2020).

BIBLIOGRAPHY

Abaschnik, V. 'Eschenmayer, Adolph Karl August.' In the *Bloomsbury Dictionary of Eighteenth-Century German Philosophers*, edited by H. F. Klemme and M. Kuehn. London: Bloomsbury, 2016. 194–5.

Amundson, R. *The Changing Role of the Embryo in Evolutionary Thought: Roots of Evo-Devo*. Cambridge: Cambridge University Press, 2005.

Autenrieth, F. F. *Verzeichniß der von Herrn Dr. v. Kielmeyer . . . hinterlassenen Bibliothek*. Stuttgart, 1845.

Ayrault, R. *La genèse du romantisme allemand*. 4 volumes. Paris: Aubier, 1961–76.

Bach, T. 'Kielmeyer als "Vater der Naturphilosophie?" Anmerkungen zu seiner Rezeption im deutschen Idealismus.' In K. T. Kanz (ed.), *Philosophie des Organischen in der Goethezeit: Studien zu Werk und Wirkung des Naturforschers Carl Friedrich Kielmeyer (1765–1844)*. Stuttgart: Franz Steiner, 1994. 232–51.

Bach, T. 'Was ist das Tierreich anders als der anatomierte Mensch . . . ? Oken in Göttingen (1805–1807).' In O. Breidbach et al. (eds), *Lorenz Oken, 1779–1851: Ein politischer Naturphilosoph*. Weimar: Böhlaus, 2001. 15–34.

Bach, T. *Biologie und Philosophie bei C. F. Kielmeyer und F. W. J. Schelling*. Stuttgart: Frommann-Holzboog, 2001.

Bach, T. '"Aber die organische Natur hat keine Geschichte . . .": Hegel und die Naturgeschichte seiner Zeit.' In R. Beuthan (ed.), *Geschichtlichkeit der Vernunft beim Jenaer Hegel*. Heidelberg: Universitätsverlag, 2006. 57–80.

Bach, T. 'Organische Physik als vergleichende Phänomenologie des Organischen. Anmerkungen zum wissenschaftshistorischen Ort von Carl Friedrich Kielmeyers System der organischen Kräfte', in O. Breidbach and R. Burwick (eds), *Physik um 1800. Kunst, Naturwissenschaft oder Philosophie?* Munich: Fink, 2012. 187–222.

Ballauf, T. and E. Scheerer. 'Organ.' In J. Ritter *et al.* (eds), *Historisches Wörterbuch der Philosophie*, vol. 6 (1984). 1317–25.

Balss, H. 'Kielmeyer als Biologe.' *Sudhoffs Archiv* 23 (1930): 268–88.

Bateson, G. *Steps to an Ecology of Mind*. Chicago: University of Chicago Press, 1972.

Beiser, F. C. *German Idealism: The Struggle against Subjectivism*. Harvard: Harvard University Press, 2001.

Berger, B. 'Schelling, Hegel, and the History of Nature.' *The Review of Metaphysics* (2020). 531–67.

Bersier, G. 'Visualising Carl Friedrich Kielmeyer's Organic Forces: Goethe's Morphology on the Threshold of Evolution.' *Monatshefte* 97.1 (2005): 18–32.

Blumenbach, J. F. *Handbuch der Naturgeschichte*. Göttingen: Dieterich, 1779/80.

Blumenbach, J. F. *Über den Bildugstrieb*. Göttingen: Dieterich, 1791.

Blumenbach, J. F. *Über den Bildungstrieb und das Zeugungsgeschäfte*. Stuttgart: Gustav Fischer, 1971.

Blumenberg, H. *Paradigms for a Metaphorology*. Translated by Robert Savage. Ithaca, NY: Cornell University Press, 2010.

Bowker, G., 'How to Be Universal: Some Cybernetic Strategies, 1943–70.' *Social Studies of Science* 23.1 (1993): 107–27.

Bowler, P. 'Theories of Generation and the History of Life.' In R. Stephenson and D. N. Wagner (eds), *The Secrets of Generation. Reproduction in the Long Eighteenth Century*. Toronto: Toronto University Press, 2015. 79–99.
Boyle, R. *Works*, 14 vols, edited by M. Hunter and E. B. Davis. London: Pickering and Chatto, 1999–2000.
Breidbach, O. 'Das Organische in Hegels Jenaer Naturphilosophie.' In K. Vieweg (ed.), *Hegels Jenaer Naturphilosophie*. Munich: Wilhelm Fink Verlag, 1998. 309–18
Breitenbach, A. 'Teleology in Biology: A Kantian Approach.' *Kant Yearbook*, 1.1 (2009): 31–56.
Buffon, G. -L. *Histoire naturelle générale et particulière*. Paris, 1749–89.
Burbidge, J. *Real Process: How Logic and Chemistry Combine in Hegel's Philosophy of Nature*. Toronto: University of Toronto Press, 1996.
Buttersack, F. 'Karl Friedrich Kielmeyer (1765–1844). Ein vergessenes Genie.' *Sudhoffs Archiv für Geschichte der Medizin* 23 (1930): 236–46.
Caneva, K. L. 'Teleology with regrets.' *Annals of Science* 47 (1990): 291–300.
Canguilhem, G. *Wissenschaftsgeschichte und Epistemologie. Gesammelte Aufsätze*. Frankfurt am Main: Suhrkamp, 1979.
Carus, C. G. *Geheimnisvoll am lichten Tag*, edited by Hans Kern. Leipzig: Reclam, 1944.
Cheung, T. 'From the Organism of a Body to the Body of an Organism: Occurrence and Meaning of the Word "Organism" from the Seventeenth to the Nineteenth Centuries.' *The British Journal for the History of Science* 39.3 (2006): 319–39.
Chiereghin, F. 'Finalità e idea della vita: la ricezione hegeliana della teleologia di Kant.' *Verifiche*, 19.1 (1990): 127–230.
Clarke, E. and L. S. Jacyna. *Nineteenth-Century Origins of Neuroscientific Concepts*. Berkeley: University of California Press, 1987.
Coleman, W. 'Kielmeyer, Carl Friedrich', in C. C. Gillipsie (ed.), *Dictionary of Scientific Biography* 7 (1973): 366–9.
Coleman, W. 'Limits of the Recapitulation Theory: Carl Friedrich Kielmeyer's Critique of the Presumed Parallelism of Earth History, Ontogeny, and the Present Order of Organisms.' *Isis* 64.3 (1973): 341–50.
Cuvier, G. *Le Ménagerie du Muséum nationale d'histoire naturelle*. Paris: Miger et Renouard, 1804.
Cuvier, G. 'Dernière leçon.' *Le Temps* (5 June 1832): 14445–54.
Derham, W. *Physico and Astro Theology: or, a Demonstration of the Being and Attributes of God*, volume 1. London, 1786.
Detienne, M. *Les dieux d'Orphée*. Paris: Gallimard, 2007.
Dougherty, F. 'Über den Einfluß Johann Friedrich Blumenbachs aufs Kielmeyers feierliche Rede von 1793.' In K. T. Kanz (ed.), *Philosophy des organischen in der Goethezeit: Studien zu Werk und Wirkung des Naturforschers Carl Friedrich Kielmeyer*. Stuttgart: Steiner, 1994. 50–80.
Drosdowski, D. (ed.). *Duden. Das große Wörterbuch der deutschen Sprache in sechs Bänden*. 6 vols. Mannheim: Duden, 1981.
Egerton, F. N. 'Changing Concepts of the Balance of Nature,' *The Quarterly Review of Biology* 48.2 (1973): 322–50.
Eggers, M. *Vergleichendes Erkennen. Zur Wissenschaftsgeschichte und Epistemologie des Vergleichs und zur Genealogie der Komparatistik*. Heidelberg: Winter, 2006.
Engels, E.-M. *Die Teleologie des Lebendigen: Kritische Überlegungen ur Neuformulierung des Teleologie-problems in der angloamerikanischen Wissenschaftstheorie*. Berlin: Dunker and Humblot, 1982.

Engels, E.-M. 'Die Lebenskraft – metaphysisches Konstrukt oder methodologisches Instrument? Überlegungen zum Status von Lebenskräften in Biologie und Medizin im Deutschland des 18. Jahrhunderts.' In K. T. Kanz (ed.), *Philosophy des organischen in der Goethezeit: Studien zu Werk und Wirkung des Naturforschers Carl Friedrich Kielmeyer*. Stuttgart: Steiner, 1994. 127–52.

Engelstein, S. 'Sexual Division and New Mythology: Goethe and Schelling.' In S. Lettow and G. Rupik (eds), *Conceiving Reproduction in German Naturphilosophie* (forthcoming).

Erlin, M. 'Reluctant Modernism: Moses Mendelssohn's Philosophy of History.' *Journal of the History of Ideas* 63.1 (2002): 83–92.

Eschenmayer, A. C. A. *Sätze aus der Natur-Metaphysik, auf chemische und medicinische Gegenstände angewendt*. Tübingen, 1797.

Eschenmayer, A. K. A. 'Spontaneity = World Soul, or On the Highest Principle of Philosophy of Nature', translated by J. Kahl and D. Whistler. In B. Berger and D. Whistler, *The Schelling-Eschenmayer Controversy, 1801: Nature and Identity*. Edinburgh: Edinburgh University Press, 2020. 17–45.

Ferrini, C. 'From Geological to Animal Nature in Hegel's Idea of Life.' *Hegel-Studien* 44 (2009): 45–94

Ferrini, C. 'Reason Observing Nature' in K. R. Westphal (ed.), *The Blackwell Guide to Hegel's Phenomenology of Spirit*. Oxford: Wiley-Blackwell, 2009. 92–135.

Fischer, F. *Die Metaphysik, von empirischem Standpunkte aus dargestellt*. Basel: Schweighäuser, 1847.

Forster, M. N. *After Herder. Philosophy of Language in the German Tradition*. Oxford: Oxford University Press, 2010.

Foucault, M. *Die Ordnung der Dinge. Eine Archäologie der Humanwissenschaften*. Frankfurt am Main: Suhrkamp, 1974.

Fusco, G. and A. Minelli. *Evolving Pathways: Key Themes in Evolutionary Developmental Biology*. Cambridge: Cambridge University Press, 2008.

Gallagher, S. *Enactivist Interventions: Rethinking the Mind*. Oxford: Oxford University Press, 2017.

Gambarotto, A. *Vital Forces, Teleology, and Organization: Philosophy of Nature and the Rise of Biology in Germany*. Dordrecht: Springer, 2018.

Gehler, J. S. T. *Physikalisches Wörterbuch oder Versuch einer Erklärung der vornehmsten Begriffe der Kunstwörter der Naturlehre*, 6 vols. Second edition. Leipzig: Schwickert, 1799.

Glass, B., O. Temkin and W. L. Straus (eds). *Forerunners of Darwin 1745–1859*. Baltimore: The Johns Hopkins Press, 1959.

Gliboff, S. *H.G. Bronn, Ernst Haeckel, and the Origins of German Darwinism*. Cambridge, MA: MIT Press, 2008.

Goethe, J. W. *Die Schriften zur Naturwissenschaft* (Leopoldina Ausgabe). 21 volumes. Edited by D. Kuhn. Weimar: Hermann Böhlau Nachfolger, 1947.

Goethe, J. W. *Werke. Hamburger Ausgabe*. 14 volumes. Edited by E. Trunz. München: Beck, 1975.

Goethe, J. W. *Schriften zur Naturwissenschaft*. Stuttgart: Reclam, 1977,

Goethe, J. W. *Scientific Studies*, Princeton: Princeton University Press, 1995.

Goethe, J. W. *Tagebücher*. Volume II.1: 1790–1800. Edited by by E. Zehm, Stuttgart: Metzler, 2000.

Görres, J. 'Prinzipien einer neuen Begründung der Gesetze des Lebens durch Dualism und Polarität (Fortsetzung).' In R. Stein (ed.), *Naturwissenschaftliche, kunst- und naturphilosophische Schriften I (1800–1803)*. Köln: Im Gilde Verlag, 1932. 45–56.

Gould, S. J. and R. Lewontin. 'The Spandrels of San Marco and the Panglossian Paradigm: A Critique of the Adaptationist Programme.' *Proceedings of the Royal Society B* 205.1161 (1971): 581–98.
Gould, S. J. *Ontogeny and Phylogeny*, Harvard: Harvard University Press, 1977.
Grant, I. H. *Philosophies of Nature after Schelling*. London: Continuum, 2006.
Grimm, J. and W. Grimm (eds). *Deutsches Wörterbuch von Jacob und Wilhelm Grimm*. Leipzig 1854–1961.
Steffens, H. *Lebenserrinerungen*. Jena: Eugen Diedrichs, 1908.
Haller, A. von. 'Vorrede' to G.-L. Buffon, *Allgemeine Historie der Natur nach allen ihren besonderen Theilen abgehandelt . . .*, Erster Theil. Hamburg und Leipzig, 1750. ix–xxii.
Haller, A. von. 'De partibus corporis humani sensibilibus et irritabilibus.' *Commentarii Societatis Regiae Scientiarum Gottingensis*, vol. 2. Göttingen, 1753.
Haller, A. von. *Anfangsgründe der Phisiologie des menschlichen Körpers*. Berlin, 1759–76.
Haller, A. von. *Elementa physiologiæ corporis humani*, volume 1. Lausanne: Marc-Michel Bousquet, 1757.
Haller, A. von. 'Von den empfindlichen und reizbaren Theilen des Menschlichen Körpers. Eine Vorlesung: Am 22ten April 1752, in der kön. Gesells. der Wiss. zu Göttingen.' In *Sammlung kleiner Hallerischer Schriften*. Bern, 1772. 1–101.
Haller, A. von. 'A Dissertation on the Sensible and Irritable Parts of Animals.' Edited by O. Temkin. *Bul. Hist. Med.* 4 (1936): 651–91.
Hegel, G. W. F. *The Science of Logic*. Translated by A. V. Miller. Amherst: Humanity Books, 1969.
Hegel, G. W. F. *Werke in 20 Bänden*. Frankfurt am Main: Suhrkamp Verlag, 1970.
Hegel, G. W. F. *Philosophy of Mind. Part III of the Encyclopaedia of the Philosophical Science*. Translated by William Wallace and A. V. Miller. Oxford: Clarendon Press, 1971.
Hegel, G. W. F. *Phenomenology of Spirit*. Translated by A. V. Miller. Oxford: Clarendon Press, 1977.
Hegel, G. W. F., *Phänomenologie des Geistes*. Edited by W. Bonsiepen and R. Heede, Hamburg: Felix Meiner, 1980.
Hegel, G. W. F. *The Encyclopaedia Logic, Part I of the Encyclopaedia of the Philosophical Sciences*. Translated by T.F. Geraets, W.A. Suchting and H. S. Harris. Indianapolis: Hackett, 1991.
Hegel, G. W. F. *Philosophy of Nature, Part II of the Encyclopaedia of the Philosophical Sciences*, Second Edition. Translated by A. V. Miller. Oxford: Oxford University Press, 2004.
Hegel, G. W. F. *Vorlesungen Über die Philosophie der Natur*. Edited by W. Bonsiepen. Hamburg: Felix Meiner, 2012.
Heims, J. *The Cybernetic Group*. Cambridge, MA: MIT Press, 1993.
Henle, F. G. J. *Handbuch der rationellen Pathologie*, 2 volumes. Braunschweig: Vieweg, 1846–53.
Herder, J. G. *Ideen zur Philosophie der Geschichte der Menschheit*, 4 volumes. Riga und Leipzig, 1784–91.
Herder, J. G. *Outlines of a Philosophy of the History of Man*. Edited and translated by T. Churchill. London: J. Johnson/L. Hansard, 1803.
Herder, J. G. *Ideen zur Philosophie der Geschichte der Menschheit*. Edited by Heinrich Luden. Leipzig: Hartnoch, 1841.
Herder, J. G. *Johann Gottfried Herder. Selected Early Works 1764–1767*. Edited by E. A. Menze and K. Menges. Pennsylvania: Pennsylvania State University Press, 1992.

Herder, J. G. *Ideen zur Philosophie der Geschichte der Menschheit*. Edited by W. Pross. Munich: Hanser, 2002.

Herder, J. G. *Philosophical Writings*. Edited by M. Forster. Cambridge: Cambridge University Press, 2002.

Heuser-Keßler, M.-L. 'Raum, Zeit, Kraft und Mannigfaltigkeit. Kant und die Forschungsmethodologie der Physik des Organischen in Kielmeyers "Rede".' In K. T. Kanz (ed.), *Philosophy des organischen in der Goethezeit: Studien zu Werk und Wirkung des Naturforschers Carl Friedrich Kielmeyer*. Stuttgart: Steiner, 1994. 111–26.

Heylighen, F. and C. Joslyn. 'Cybernetics and Second-Order Cybernetics.' In R. A. Meyers (ed.), *Encyclopedia of Physical Science & Technology*. New York: Academic Press, 2001.

Hogrebe, W. *Metaphysik und Mantik. Die Deutungsnatur des Menschen (Système orphique de Iéna)*. Frankfurt: Suhrkamp, 1992.

Holland, J. *German Romanticism and Science. The Procreative Politics of Goethe, Novalis and Ritter*. London: Routledge, 2009.

Honegger, C. *Die Ordnung der Geschlechter. Die Wissenschaften von Menschen und das Weib*. Frankfurt am Main: Campus, 1991.

Hopwood, N. 'The Keywords "Generation" and "Reproduction".' In N. Hopwood et al. (eds). *Reproduction from Antiquity to the Present*. Cambridge: Cambridge University Press, 2018. 287–304.

Horstmann, R. -P. 'Maimon's criticism of Reinhold's *Satz des Bewusstseins*.' In L. W. Beck (ed.), *Proceedings of the Third International Kant-Congress*. Dordrecht: Reidel, 1972.

Houlgate, S. *Hegel's Phenomenology of Spirit: A Reader's Guide*. London: Bloomsbury, 2012.

Hübscher, A. 'Kleine Nachlese.' *Schopenhauer-Jahrbuch* (1983): 154–64.

Humboldt, A. von and A. Bonpland. *Beobachtungen aus der Zoologie und vergleichenden Anatomie*. Tübingen, 1806–9.

Humboldt. A. von. *Essay on the Geography of Plants*. Translated by S. Romanowski. Chicago: Chicago University Press, 2009.

Huneman, P. 'Assessing the Prospects for a Return of Organisms in Evolutionary Biology.' *History and Philosophy of the Life Sciences*, 32.2-3 (2010): 341–71.

Huneman, P. 'Kant's Concept of Organism Revisited: A Framework for a Possible Synthesis between Developmentalism and Adaptationism?' *The Monist*, 100.3 (2017): 373–90.

Hyppolite, J. *Genesis and Structure in Hegel's Phenomenology of Spirit*. Translated by S. Cherniak and J. Heckman. Evanston: Northwestern University Press, 1974.

Jaeger, G. F. 'Ehrengedächtniss des königl. württembergischen Staatsraths von Kielmeyer', *Novorum Actorum Academiae Caesareae Leopoldino-Carolinae Naturae Curiosorum* 21.2 (1845): xvii–xcii.

Jahn, I. *Grundzüge der Biologiegeschichte*. Jena: Fischer, 1990.

Jahn, I. 'War Gustav Wilhelm Münter (1804–1870) ein "Plagiator" Kielmeyers?' In K. T. Kanz (ed.), *Philosophie des Organischen in der Goethezeit*. Stuttgart: Steiner, 1994. 174–210.

Jantzen, J. 'Eschenmayer und Schelling. Die Philosophie in ihrem Übergang zur Nichtphilosophie.' In W. Jaeschke (ed.), *Religionsphilosophie und spekulative Theologie: Die Streit um die Göttliche Dinge (1799–1812)*. Hamburg: Felix Meiner, 1994. 74–97.

Jantzen, J. 'Physiologische Theorien.' In M. Durner et al. (eds), *Wissenschaftshistorischer Bericht zu Schellings naturphilosophischen Schriften 1797–1800*. Stuttgart: Frommann-Holzboog, 1994. 373–668.

Jantzen, J. 'Adolph Karl August von Eschenmayer.' In T. Bach and O. Breidbach (eds), *Naturphilosophie nach Schelling*. Stuttgart: Frommann-Holzboog, 2005. 153–80.

Jardine, N. 'The significance of Schelling's "Epoch of a wholly new natural history". An essay on the realization of questions.' In R. S. Woolhouse (ed.), *Metaphysics and philosophy of science in the seventeenth and eighteenth centuries*. Dordrecht: Reidel, 1988. 327-50.

Jardine, N. *The Scenes of Inquiry: On the Reality of Questions in the Science*. Oxford: Clarendon Press, 1991.

Kant, I. *Allgemeine Naturgeschichte und Theorie des Himmels, oder Versuch von der Verfassung und dem mechanischen Ursprunge des ganzen Weltgebäudes nach Newtonischen Grundsätzen abgehandelt*. Fourth edition. Zeitz, 1808.

Kant, I. *Critique of Pure Reason*. Translated by P. Guyer and A. Wood. Cambridge: Cambridge University Press, 1999.

Kant, I. *Critique of the Power of Judgment*. Translated by P. Guyer. Cambridge: Cambridge University Press, 2000.

Kant, I. *Anthropology, History, Education*. Edited by G. Zöller and R. B. Louden. Cambridge: Cambridge University Press, 2007.

Kant, I. *Critique of Judgement*. Translated by James Creed Meredith. Oxford: Oxford University Press, 2008.

Kant, I. *Critique of Judgment*. Translated by W. S. Pluhar. Indianapolis, IN: Hackett, 1987.

Kanz, K. T. (1989), 'Carl Friedrich Kielmeyer, Lichtenberg und Göttingen 1786-1796.' *Lichtenberg-Jahrbuch* (1989): 140-60.

Kanz, K. T. *Kielmeyer-Bibliographie: Verzeichnis der Literatur von und über den Naturforscher Carl Friedrich Kielmeyer (1765-1844)*. Stuttgart: GNT Verlag, 1991.

Kanz, K. T. 'Carl Friedrich Kielmeyer (1765-1844): Wegbereiter des Entwicklungsgedankens.' In H. Albrecht (ed.), *Schwäbische Forscher und Gelehrte*. Stuttgart: DRW-Verlag, 1992. 72-6.

Kanz, K. T. 'Einführung.' In C. F. Kielmeyer,: *Über die Verhältniße der organischen Kräfte unter einander in der Reihe der verschiedenen Organisationen, die Gesetze und Folgen dieser Verhältniße*. Marburg an der Lahn: Basilisken-Presse, 1993. 9-70.

Kanz, K. T. 'Die Naturgeschichte (Botanik, Zoologie, Mineralogie) an der Hohen Karlsschule in Stuttgart (1772-1794).' *Jahreshefte der Gesellschaft für Naturkunde in Württemberg* 148 (1993): 5-23.

Kanz, K. T. (ed.). *Philosophie des Organischen in der Goethezeit. Studien zu Werk und Wirkung des Naturforschers Carl Friedrich Kielmeyer (1765-1844)*. Stuttgart: Steiner, 1994.

Kanz, K. T. 'Carl Friedrich Kielmeyer (1765-1844) – Leben, Werk, Wirkung: Perspektiven der Forschung und Edition.' In K. T. Kanz (ed.), *Philosophie des organischen in der Goethezeit: Studien zu Werk und Wirkung des Naturforschers Carl Friedrich Kielmeyer*. Stuttgart: Steiner, 1994. 13-32.

Kanz, K. T. '"Wie einzelne Menschen, so haben auch einzelne Entdeckungen ihr besonderes Glück." Gustav Wilhelm Münter (1804-1870) und seine Entdeckung der ‚Physik der organischen Körper' oder der wissenschaftliche Größenwahn eines anatomischen Kustos.' In J. Schulz (ed.), *Fokus Biologiegeschichte. Zum 80. Geburtstag der Biologiehistorikerin Ilse Jahn*. Berlin: Akadras, 2002. 89-110.

Kanz, K. T. '"Halle zum vollkommnen Muster einer Academie zu organisieren." Johann Christian Reils Berufungskorrespondenz mit J. H. F. Autenrieth und C. F. Kielmeyer in Tübingen.' In F. Steger (ed.), *Johann Christian Reil. Universalmediziner, Stadtphysikus, Wegbereiter von Psychiatrie und Neurologie*. Gießen: Psychosozial-Verlag, 2014. 37-68.

Kanz, K. T. '"Die Besten der Gelehrtenwelt beklagen Ihr Stillschweigen": Schellings Einladung an Carl Friedrich Kielmeyer zur Mitarbeit an den *Jahrbüchern der Medicin als Wissenschaft*.' *Schelling-Studien* 5 (2017): 183-96.

Keularz, J. 'Rolston.' In W. B. Drees (ed.), *Is Nature Ever Evil? Religion, Science, and Value.* London: Routledge, 2003.

Kielmeyer, C. F. 'Gutachten des Herrn Staatsraths von Kielmeyer über die Lapostolle'sche Schrift, betreffend Blitz- und Hagelableiter aus Strohseilen.' *Correspondenzblatt des Würtembergischen Landwirthschaftlichen Vereins* 7 (1825): 268–76.

Kielmeyer, C. F. *Natur und Kraft. Gesammelte Schriften.* Edited by F.-H. Holler. Berlin: Keiper, 1938.

Kielmeyer, C. F. *Über die Verhältnisse der organischen Kräfte untereinander, in der Reihe der verschiedenen Organisationen, die Gesetze und Folgen diese Verhältnisse.* Edited by K. T. Kanz. Marburg: Basiliken-Presse, 1993.

Kielmeyer, C. F. and G. Jäger (eds). *Amtlicher Bericht über die Versammlung deutscher Naturforscher und Ärzte zu Stuttgart.* Stuttgart: Mezler, 1835.

Kisner, W. *Ecological Ethics and Living Subjectivity in Hegel's Logic: The Middle Voice of Autopoietic Life.* Hampshire: Palgrave Macmillan, 2014.

Klein, U. and W. Lefèvre. *Materials in Eighteenth-Century Science: A Historical Ontology.* Cambridge, MA: MIT Press, 2007.

Köpke, R. *Die Gründung der Königlichen Friedrich-Wilhelms-Universität zu Berlin.* Berlin: Schade, 1860.

Kuhn, D. 'Uhrwerk oder Organismus. Karl Friedrich Kielmeyers System der organischen Kräfte.' *Nova Acta Leopoldina*, Neue Folge 36 (1970): 157–67.

Kuhn, D. 'Der naturwissenschaftliche Unterricht an der Hohen Karlsschule.' *Medizinhistorisches Journal* 11 (1976): 319–34.

Kuhn, D. 'Uhrwerk oder Organismus: Carl Friedrich Kielmeyers System der organischen Kräfte.' In K. T. Kanz (ed.), *Philosophy des organischen in der Goethezeit: Studien zu Werk und Wirkung des Naturforschers Carl Friedrich Kielmeyer.* Stuttgart: Steiner, 1994. 33–49.

Landwehr, A. *Geschichte des Sagbaren. Einführung in die Historische Diskursanalyse.* Tübingen: Diskord, 2001.

Larson, J. 'Vital Forces: Regulative Principles or Constitutive Agents? A Strategy in German Physiology, 1786–1802.' *Isis* 70.2 (1979): 235–49.

Lenoir T. 'The Göttingen School and the Development of Transcendental *Naturphilosophie* in the Romantic Era.' *Studies in the History of Biology* 5 (1978): 111–205.

Lenoir, T. 'Kant, Blumenbach, and Vital Materialism in German Biology.' *Isis* 71 (1980): 77–108.

Lenoir, T. 'The Göttingen School and the Development of Transcendental *Naturphilosophie* in the Romantic Era.' *Studies in History of Biology* 5 (1981): 111–205.

Lenoir, T. 'Teleology without Regrets. The Transformation of Physiology in Germany: 1790–1847.' *Studies in History and Philosophy of Science, Part A*, 12.4 (1981): 293–354.

Lenoir, T. *The Strategy of Life: Teleology and Mechanics in Nineteenth-Century Biology.* Dordrecht: Reidel, 1982.

Lenoir, T. 'Kant, Von Baer, and Causal-Historical Thinking in Biology.' *Poetics Today* 9.1 (1988): 103–15.

Lepenies, W. *Das Ende der Naturgeschichte. Wandel kultureller Selbstverständlichkeiten in den Wissenschaften des 18. und 19. Jahrhunderts.* Frankfurt am Main: Suhrkamp, 1978.

Lettow, S. 'Generation, Genealogy, and Time: The Concept of Reproduction from Histoire naturelle to Naturphilosophie.' In S. Lettow (ed.), *Reproduction, Race and Gender in Philosophy and the Early Life Sciences.* Albany, NY: SUNY, 2014. 21–44.

Lind, G. *Physik im Lehrbuch 1700–1850.* Berlin: Springer, 1992.

Lippmann, E. *Urzeugung und Lebenskraft: Zur Geschichte dieser Probleme von den ältesten Zeiten an bis zu den Angängen des 20. Jahrhundrets.* Berlin: Julius Springer, 1933.

Locke, J. *An Essay Concerning Human Understanding*. Oxford: Clarendon, 1975.
Love, A. *Conceptual Change in Biology: Scientific and Philosophical Perspectives on Evolution and Development*. Dordrecht: Springer, 2015.
Lovejoy, A. O. *The Great Chain of Being: A Study of the History of an Idea*. Cambridge, MA: Harvard University Press, 1971.
Löw, R. *Pflanzenchemie zwischen Lavoisier und Liebig*. Munich: Donau-Verlag, 1977.
Martius, C. F. P. von. 'Denkrede auf Carl Friedrich von Kielmeyer.' In C. F. P. von Martius, *Akademische Denkreden*. Leipzig, 1866. 181–209.
Mayer, A. K. 'Eine Reliquie von C. Frd. Kielmeyer', *Archiv der Heilkunde* 5 (1864): 354–67; 'Nachtrag', *Archiv der Heilkunde* 6 (1865): 474–5.
Meckel, J. F. *Archiv Für Anatomie und Physiologie*. Leipzig, 1826-32.
Morgan, T. H. *Evolution and adaptation*, New York: Macmillan, 1903.
Mossio, M. et al. 'An Organizational Account of Biological Functions.' *British Journal for the Philosophy of Science*, 60.4 (2009): 813–41.
Mossio, M. and L. Bich. 'What Makes Biological Organization Teleological?' *Synthese* 194.4 (2017): 1089–1114.
Müller, G. B. 'Why an Extended Evolutionary Synthesis is Necessary.' *Interface Focus* 7.5 (2017): 1–11.
Müller, J. *Zur vergleichenden Physiologie des Gesichtssinnes des Menschen und der Thiere*. Leipzig: Cnobloch, 1826.
Müller, O. L. *Mehr Licht! Goethe mit Newton im Streit um die Farben*. Berlin: Fischer, 2015.
Müller-Wille, S. and H.-J. Rheinberger. 'Heredity –The invention of an epistemic space.' In S. Müller-Wille and H.-J. Rheinberger (eds), *Heredity Produced: At the Crossroads of Biology, Politics, and Culture, 1500–1870*. Cambridge, MA: MIT Press, 2007. 3–34.
Münter, G. W. *Allgemeine Zoologie oder Physik der organischen Körper*. Halle: Schwetschke, 1840.
Novalis. *Schriften*. 4 volumes. Edited by Paul Kluckhohn and Richard Samuel. Stuttgart: W. Kohlhammer, 1983.
Oken, L. *Abriß des Systems der Biologie*. Göttingen, 1805.
Oken, L. *Die Zeugung*. Weimar: Hermann Böhlaus Nachfahren, 2007.
Oken, L. *Gesammelte Werke*, 4 volumes. Edited by Thomas Bach et al. Weimar: Böhlaus, 2007.
Pfaff, C. H. *Über tierische Elektricität und Reizbarkeit*. Leipzig: Crusius, 1795.
Pfaff, C. H. *Lebenserinnerungen*. Kiel: Schwers'sche Buchhandlung, 1854.
Pigliucci, M. and G. Müller. *Evolution: The Extended Synthesis*. Cambridge, MA: MIT Press, 2010.
Plato. *Timaeus*. Translated by R.G. Bury. Cambridge, MA: Harvard University Press, 1975.
[Pölitz, K. H. L.] *Die Philosophie unseres Zeitalters in der Kinderkappe; von einem Manne, der auch lange in dieser Kappe gelaufen ist*. Pirna: Arnold & Pinther, 1800.
Pross, W. 'Herders Konzept der organischen Kräfte und die Wirkung der Ideen zur Philosophie der Geschichte der Menschheit *auf Carl Friedrich Kielmeyer.*' In K. T. Kanz (ed.), *Philosophy des organischen in der Goethezeit: Studien zu Werk und Wirkung des Naturforschers Carl Friedrich Kielmeyer*. Stuttgart: Steiner, 1994. 81–99.
Rauther, M. 'Ungenutzte Quellen zur Kenntnis K. F. Kielmeyer's.' *Besondere Beilage des Staats-Anzeigers für Württemberg* 6 (14 May 1921): 113–22.
Rauther, M. 'Carl Friedrich Kielmeyer zu Ehren', *Jahreshefte des Vereins für vaterländische Naturkunde in Württemberg* 94 (1938): xxiv–xxix.
Reill, P. H. *Vitalizing Nature in the Enlightenment*. Berkeley: University of California Press, 2005.

Reill, P. H. 'The Scientific Construction of Gender and Generation in the German Late Enlightenment and in German Romantic *Naturphilosophie.*' In S. Lettow (ed.), *Reproduction, Race and Gender. Philosophy and the Early Life Sciences in Context,* Albany, NY: SUNY Press, 2005. 65-82.

Reinhold, K. L. *Versuch einer neuen Theorie des menschlichen Vorstellungsvermögens.* Prague-Jena: Wiedemann u. Mauke, 1789.

Rey, R. 'La recapitulation chez les physiologists et naturalists allemands de la fin du XVIIIe et du début du XIXe siècle.' In P. Mengal (ed.), *Histoire du concept de recapitulation.* Paris: Masson, 1993. 39-54.

Rey, R., *Naissance et développement du vitalisme en France de la deuxième moitié du 18e siècle à la fin du Premier Empire.* Oxford: Voltaire Foundation, 2000.

Rheinberger, H.-J. 'Buffon: Zeit, Veränderung und Geschichte.' *History and Philosophy of the Life Sciences* 12 (1990): 203-23.

Richards, R. J. *The Meaning of Evolution: The Morphological Construction and Ideological Reconstruction of Darwin's Theory.* Chicago: University of Chicago Press, 1992.

Richards, R. J. 'Kant and Blumenbach on the *Bildungstrieb*: A Historical Misunderstanding.' *Stud. Hist. Phil. Biol. Biomed. Sci.* 31.1 (2000): 11-32.

Richards, R. J. *The Romantic Conception of Life: Science and Philosophy in the Age of Goethe.* Chicago: University of Chicago Press, 2002.

Ritter, J. W. *Beweis, dass ein beständiger Galvanismus den Lebensprozess in dem Thierreich begleite.* Weimar: Industrie-Comptoire, 1798.

Roe, S. A. *Matter, Life, and Generation Eighteenth-Century Embryology and the Haller-Wolff Debate.* Cambridge: Cambridge University Press, 1981.

Roe, S. A. 'Anatomia animata. The Newtonian physiology of Albrecht von Haller. In E. Mendelsohn (ed.), *Transformation and Tradition in the Sciences. Essays in Honour of I. Bernard Cohen.* Cambridge: Cambridge University Press, 1984. 273-300.

Rosenblueth, A., N. Wiener and J. Bigelow. 'Behavior, Purpose and Teleology.' *Philosophy of Science,* 10 (1943): 18-24.

Russell, E. S. *Form and Function. A Contribution to the History of Animal Morphology.* Chicago: University of Chicago Press, 1982.

Sandford, S. *Plato and Sex.* London: Polity Press, 2010.

Schelling, F. W. J. *Von der Weltseele: Eine Hypothese der höhern Physik zur Eklärung des allgemeinen Organismus,* Hamburg: Perthes, 1798.

Schelling, F. W. J. *Erster Entwurf eines Systems der Naturphilosophie zum Behuf seiner Vorlesungen,* Jena: Gabler, 1799.

Schelling, F. W. J. *Sämmtliche Werke,* 15 volumes. Stuttgart and Augsberg: Cotta, 1856-61.

Schelling, F. W. J. *Aus Schellings Leben,* 2 volumes. Edited by G. L. Plitt. Leipzig: S. Hirzel, 1869.

Schelling, F. W. J. *System of Transcendental Idealism.* Translated by Peter Heath. Charlottesville: University of Virginia Press, 1978.

Schelling, F. W. J. *Von der Weltseele – Eine Hypothese der Höhern Physik zur Erklärung des allgemeinen Organismus.* Stuttgart: Frommann-Holzboog, 2000/2006.

Schelling, F. W. J. *Erster Entwurf eines Systems der Naturphilosophie.* Stuttgart: Frommann-Holzboog, 2000.

Schelling, F. W. J. *First Outline of a System of the Philosophy of Nature,* trans. Keith R. Peterson (Albany, NY: SUNY Press, 2001

Schelling, F. W. J. *Historisch-kritische Ausgabe,* vol. III/1. Edited by Irmgard Möller and Walter Schieche. Stuttgart-Bad Canstatt: Frommann-Holzboog, 2001.

Schelling, F. W. J. *First Outline of a System of the Philosophy of Nature*. Translated by Keith R. Peterson. Albany: SUNY Press, 2004.

Schelling, F. W. J. *Philosophical Investigations into the Essence of Human Freedom*. Translated by J. Love and J. Schmidt. Albany: SUNY Press, 2006.

Schiebinger, L. *The Mind has No Sex? Women in the Origins of Modern Science*. Cambridge, MA: Harvard University Press, 1991.

Schlegel, F. *Kritische Friedrich-Schlegel-Ausgabe*, 35 volumes. Edited by Ernst Behler. Munich: Schöningh, 1958-.

Schlüter, H. *Die Wissenschaften vom Leben zwischen Physik und Metaphysik. Auf der Suche nach dem Newton der Biologie im 19. Jahrhundert*. Weinheim: VCH, 1985.

Schmitt, S. (ed.). *Les forces vitales et leur distrubution dans la nature. Un essai de 'systématique physiologique*, London : Brepols 2006.

Schmitt, S. 'Succession of Functions and Classification in post-Kantian Naturphilosophie around 1800'. In P. Huneman (ed.), *Understanding Purpose. Kant and the Philosophy of Biology*. Rochester: University of Rochester Press, 2007.

Schopenhauer, A. *Der handschriftliche Nachlaß*, 5 volumes. Edited by Arthur Hübscher. Munich: Deutscher Taschenbuch Verlag, 1985.

Schumacher, I. 'Karl Friedrich Kielmeyer, ein Wegbereiter neuer Ideen. Der Einfluß seiner Methode des Vergleichens auf die Biologie seiner Zeit.' *Medzinhistorisches Journal* 14 (1979): 81–99.

Siegel, C. *Geschichte der deutschen Naturphilosophie*, Leipzig: Akademische Verlagsgesellschaft, 1913.

Steigerwald, J. *Experimenting at the Boundaries of Life: Vitality in Germany around 1800*. Pittsburgh: University of Pittsburgh Press, 2019.

Steinke, H. et al. (eds) *Albrecht von Haller. Leben – Werk – Epoche*. Second edition. Göttingen: Wallstein, 2009.

Stern, R. *Hegel and the Phenomenology of Spirit*. London: Routledge, 2002.

Stichweh, R. *Zur Entstehung des modernen Systems wissenschaftlicher Disziplinen: Physik in Deutschland 1740–1890*. Frankfurt am Main: Suhrkamp, 1984.

Stone, A. 'Sexual Polarity in Schelling and Hegel.' In S. Lettow (ed.), *Reproduction, Race and Gender. Philosophy and the Early Life Sciences in Context*. Albany, NY: SUNY Press, 2014. 259–282.

Stuhlhofer, F. *Lohn und Strafe in der Wissenschaft. Naturforscher im Urteil der Geschichte*. Vienna: Böhlau, 1987.

Tetens, J. N. *Philosophische Versuche über die menschliche Natur und ihre Entwickelung*, 2 volumes. Leipzig, 1777.

Toellner, R. *Albrecht von Haller. Über die Einheit im Denken des letzten Universalgelehrten*. Wiesbaden: Steiner, 1971.

Toepfer, G. *Historisches Wörterbuch der Biologie. Geschichte und Theorie der biologischen Grundbegriffe*, 3 volumes. Stuttgart: Springer, 2001/2011.

Toepfer, G. '"Organisation" und "Organismus" – von der Gliederung zur Lebendigkeit – und zurück? Die Karriere einer Wortfamilie seit dem 17. Jahrhundert.' In M. Eggers and M. Rothe (eds), *Wissenschaftsgeschichte als Begriffsgeschichte. Terminologische Umbrüche im Entstehungsprozess der modernen Wissenschaften*. Bielefeld: de Gruyter, 2009. 83–106.

Van den Berg, H. 'The Wolffian Roots of Kant's Teleology.' *Stud Hist Biol Biomed Sci* 44 (2013): 724–34.

Vienne, F. 'Organic Molecules, Parasites, Urthiere. The Controversial Nature of Spermatic Animals, 1749–1841.' In S. Lettow (ed.), *Reproduction, Race and Gender. Philosophy and the Early Life Sciences in Context*. Albany, NY: SUNY Press, 2014. 45–63.

Voigt, J. H. 'Lehre zwischen Politik und Wirtschaft 1829–1864. Von der Real- und Gewerbeschule zur Polytechnischen Schule.' In *Festschrift zum 150jährigen Bestehen der Universität Stuttgart. Beiträge zur Geschichte der Universität* Stuttgart: Deutsche Verlags-Anstalt, 1979. 13–138.

von Wülfingen, B. B. et al. 'Temporalities of Reproduction: Practices and concepts from the eighteenth to the early twenty-first century. Introduction.' *History and Philosophy of the Life Sciences* 37.1 (2015): 1–16.

Wagner, R. *Samuel Thomas Soemmerrings Leben und Verkehr mit seinen Zeitgenossen*, 2 volumes. Leipzig: Voss, 1844.

Weber, A. and F. Varela. 'Life after Kant: Natural purposes and the autopoietic foundations of biological individuality.' *Phenomenology and the Cognitive Sciences* 1 (2002): 97–125.

Whistler, D. 'In the Hope of a Philosopher of Nature.' In A. Ezekiel and K. Mihaylova (eds), *Hope and the Limits of the Self in Classical German Philosophy*. Berlin: De Gruyter, 2021 forthcoming.

Whitehead, A. N. *Science and the Modern World*. Cambridge: Cambridge University Press, 1926.

Wiener, N. *Cybernetics: Or Control and Communication in the Animal and the Machine*. Boston, MA: MIT Press, 1948.

Wolf, J. H. *Der Begriff 'Organ' in der Medizin. Grundzüge der Geschichte seiner Entwicklung*. Munich: Fitsch, 1971.

Wolfe, C. 'On the Role of Newtonian Analogies in Eighteenth-Century Life Science.' In Z. Biener and E. Schliesser (eds), *Newton and Empiricism*. Oxford: Oxford University Press, 2014. 223–61.

Wootton, D. *The Invention of Science. A New History of the Scientific Revolution*. London: Penguin, 2015.

Zammito, J. H. 'Teleology then and now: The question of Kant's relevance for contemporary controversies over functions in biology.' *Studies in History and Philosophy of Biological and Biomedical Sciences*, 37.4 (2006): 748–70.

Zammito, J. H. *The Gestation of German Biology: Philosophy and Physiology from Stahl to Schelling*. Chicago: University of Chicago Press, 2017.

Ziche, Paul. *Mathematische und naturwissenschaftliche Modelle in der Philosophie Schellings und Hegels*. Stuttgart: Frommann-Holzboog, 1996.

INDEX

activity 69–71, 158, 167, 5, 32, 126, 214
analogy 22, 58, 85, 75, 84, 84–5, 86–7, 156 n.2, 162
attraction 38, 72, 92, 106, 161, 195

balance 37, 43, 59, 91, 94, 99–100, 110, 115–18, 121–3, 127–8, 129 n.8, 142, 152, 159, 171, 175, 188–93, 196, 198–9
biology 1–2, 19–23, 99–102, 108, 150, 156–67, 180–1
Blumenbach, J. 2, 6, 8, 12–13, 20–1, 82, 85, 92, 95, 120, 160–4, 167, 171, 173

chemical 6, 10 n.24, 58, 72, 75, 84, 127, 153, 161, 187–9, 191–7, 199, 216
comparative anatomy 9 n.13, 14, 15, 21, 49, 76, 77 n.15, 81, 88, 214–15
comparative zoology 108, 129 n.1, 162
Cuvier, G. 8, 12, 14, 22, 69–77, 89, 149, 137, 142, 149, 164, 166, 180, 181, 214, 219 n.48
cybernetics 149–66, 169 n.42

Derham, W. 117–18, 121, 127
drive 54, 59, 85, 92, 161, 170 n.55, 171, 173

economy 95, 115–17, 121, 128, 171, 176, 179
electricity 21, 36, 70, 75, 102, 181, 189
emergence 5, 42, 54–5, 94, 128, 134, 142–3, 174
empirical psychology 137, 146 n.24
empirical science 22, 97 n.37, 104, 166, 208, 209, 212–13, 217 n.12, 212–13, 216, 219 n.48
empiricism 99, 208, 212, 219 n.48
equilibrium *see* 'balance'
Eschenmayer, A. K. A. 3–4, 8, 14, 183, 187–99, 200 n.2
evolution 18, 20, 22, 60, 149–50, 155–7

extinction 6, 115–16, 123–6, 128–31, 152, 197

Fichte, J. G. 76 n.5, 77 n.6, 137, 189
function 4, 21, 85–7, 118, 120, 124, 127, 142, 153, 156, 159–60, 165–6, 168 n.22, 169 n.34, 170 n.56, 174, 176, 218 n.35

gender 172, 175–8, 180–1
Goethe, J. W. 3, 17, 21, 81, 96 n.3, 100, 134, 135, 144 n.2, 145 n.14, 172, 174–5, 177, 182–3 n.10, 183

Haller, A. 48 n.11, 84, 104–6, 160
harmony 19, 30, 70–1, 89, 117, 120–1, 136–7
Hegel, G. W. F. 151, 153, 156–7, 166, 172, 176–7, 194, 203–16

irritability 4, 21, 32, 35–7, 41–3, 84–5, 92–4, 105–6, 108–10, 118–21, 164–5, 170 n.55, 177, 189, 205–15

Kant, I. 1, 3, 69, 72–5, 82, 85–6, 90–1, 95, 137, 153–7, 159, 163, 165, 171, 179, 194

Leibniz, G. W. 70–1, 120
Lesage, G. L. 46, 49 n.24, 75
lever 188–94, 196, 199, 200 n.6

machine 5, 30, 47, 53, 55, 89, 92, 95, 99, 109, 133, 158, 162–3, 172
matter 59, 60, 72–5, 82–5, 87, 89–90, 92–3, 104, 118, 141, 154, 161, 179–80, 184 n.26, 194–6
mechanics 61, 82, 85, 93, 102, 104, 106, 156, 196, 187–9
mechanism 1, 8, 55, 61, 82–9, 93, 95–6, 105–6, 153–9, 161–2, 164, 166, 187–99, 194
mind 70–1, 83, 86, 136–7, 140, 146 n.22, 154, 167, 169 n.42, 198, 204

motion 31, 45, 55, 62, 84, 92, 106, 108, 110, 139, 150, 154, 160, 162, 188–94, 196–9

natural history 1, 3, 6, 12, 13, 15–17, 20, 22, 53–63, 76 n.1, 76 n.3, 85–9, 92, 99, 101–4, 112 n.36, 122, 125, 127, 133–6, 138–44, 151–4, 159–60, 162, 179–81, 215
Naturphilosophie 3–4, 8, 23, 133, 149–51, 154, 160, 166, 172, 180, 181, 188–9, 191, 193–4, 199, 203–16, 217 n.12, 229 n.43
Newtonianism 82–7, 92, 95, 96 n.12, 102

Oken, L. 135, 149, 172, 176–7
ontogeny 6, 20, 133–44, 149–50, 159, 166
organ 5, 18, 30, 32–4, 36, 46, 118, 124, 150, 160, 163–4, 173–7, 198
organisation 5, 9 n.19, 31–8, 41–54, 88–94, 118–19, 162
organism 108–9, 118–20, 155–65, 171, 212–13

philosophy of nature *see* '*Naturphilosophie*'
phylogeny 6, 20, 149, 166, 181 n.1
physics 7, 13, 15, 19, 55, 58, 85, 88, 91, 94–5, 99–100, 101–5, 108–11, 162–3, 191, 212
physiology 3–4, 12–13, 17, 19, 21, 54, 82, 84–5, 104–5, 107, 159
plant 4, 15, 18, 32–9, 41–4, 56–9, 62, 66–7, 71, 76, 83, 93–4, 107, 110, 139, 164–5, 174, 176–7, 210, 215
Plato 71, 154, 179–80
polarity 75, 93, 174–5, 180–1, 196
purpose, purposiveness 87, 153–9, 163–7

race 8, 10 n.27, 47, 57, 185 n.39
recapitulation 6, 8, 20, 22, 133–9, 142–4, 149, 166

reproduction 4, 21, 39–41, 85, 92–4, 106, 117–18, 120–1, 159, 161, 164–5, 173–7, 182 n.3, 182 n.7, 184 n.28, 200 n.3, 205–15, 218 n.38

Schelling, F. W. J. 1, 3, 21, 67, 81, 120, 134, 138, 140–4, 145 n.6, 147 n.29, 153–4, 159, 165–9, 173–4, 177, 179–81, 188, 195, 205, 215–16
sensibility 4, 21, 33–7, 41, 43, 84–5, 94, 108–10, 118, 120–1, 159–60, 162, 164–5, 171, 177, 205–15
series of organisations 5, 6, 18, 29, 31, 33, 35–7, 41–4, 65, 66–7, 88, 91, 93, 107–10, 119–20, 122–3, 150, 159, 162, 164–5, 173, 176, 178
sex 171–86, 185, 211
Spallanzani, L. 40, 48 n.17
species 4, 7, 30–47, 56–7, 63 n.5, 67, 89–91, 94–5, 109–10, 116–28, 138, 144, 165, 172, 176, 198–9, 210, 204–7, 214–18, 220–1, 224–6

technology 128
teleology 85, 93, 101, 127, 151, 154–8, 162–6
Trembley, A. 48, 184

understanding, the 6, 32, 45, 83, 87, 88, 90, 151, 185 n.52, 204, 208, 219 n.43
universal 32, 33, 38, 39, 53, 55, 61, 72–5, 90, 92, 169 n.42, 171, 172, 204, 210, 212, 214, 215
universal organism 145 n.15, 153, 159, 162, 164–6

Windischmann, K. J. H. 7–8, 65, 667 n.1, 131 n.25, 133–4, 178–9, 185 n.41
Wolff, C. F. 157, 160–1
wonder 47, 59, 121
world soul *see* 'universal organism'

www.ingramcontent.com/pod-product-compliance
Lightning Source LLC
Chambersburg PA
CBHW072148290426
44111CB00012B/2004